FOUNDATIONS IN PROBLEM SOILS

FOUNDATIONS IN PROBLEM SOILS

A Guide to Lightly Loaded Foundation Construction for Challenging Soil and Site Conditions

Steven J. Greenfield
University of California, Davis

C. K. Shen
University of California, Davis

Edited by:
S. Scot Litke

ADSC:
The International Association
of Foundation Drilling
P.O. Box 280379
Dallas, TX 75228

PRENTICE HALL, Englewood Cliffs, NJ 07632

Library of Congress Cataloging-in-Publication Data

Greenfield, Steven J.
Foundations in problem soils : a guide to lightly loaded foundation construction for challenging soil and site conditions / by Steven J. Greenfield, C. K. Shen : edited by S. Scot Litke.
p. cm.
"In cooperation with ADSC: the International Association of Foundation Drilling."
Includes bibliographical references and index.
ISBN 0-13-318908-2. — ISBN 0-13-318890-6 (pbk.)
1. Foundations—Design and construction. 2. Soil mechanics.
I. Shen, Chih-Kang II. Litke, S. Scot (Stephen Scot)
III. Association of Drilled Shaft Contractors (U.S.) IV. Title.
TA775.G675 1992
624.1'6—dc20 91-23546
CIP

Acquisitions editor: ***Doug Humphrey***
Editorial/production supervision
and interior design: ***Richard DeLorenzo***
Copy editor: ***James Tully***
Cover design: ***Bruce Kenselaar***
Cover photo: ***Mark Elliott/Anderson Drilling; Lakeside, California***
Prepress buyer: ***Linda Behrens***
Manufacturing buyer: ***David Dickey***
Editorial assistant: ***Jaime Zampino***

Printed in the United States of America

10 9 8 7 6 5 4 3 2 1

ISBN 0-13-318908-2
ISBN 0-13-318890-6 (pbk.)

Prentice-Hall International (UK) Limited, *London*
Prentice-Hall of Australia Pty. Limited, *Sydney*
Prentice-Hall Canada Inc., *Toronto*
Prentice-Hall Hispanoamericana, S.A., *Mexico*
Prentice-Hall of India Private Limited, *New Delhi*
Prentice-Hall of Japan, Inc., *Tokyo*
Simon & Schuster Asia Pte. Ltd., *Singapore*
Editora Prentice-Hall do Brasil, Ltda., *Rio de Janeiro*

Dedicated with love
to my wife Teri
and my parents

Contents

Acknowledgments

I would like to thank all of the contractors, architects, and engineers I interviewed from across the United States who took the time to share their knowledge and experience. Their insights and observations are what make up the heart of this guide. I am especially grateful to those who invited me to spend some time observing one of their construction projects.

I would also like to thank Mr. Herb Minatre, president of Bayshore Systems, Incorporated, Benicia, California, who was the impetus and driving force behind this endeavor. Herb's enthusiasm, friendship, direction, and many review hours were critical to the guide's completion.

I am grateful to Dr. C. K. Shen, Department Chairman of Civil Engineering at the University of California at Davis, as co-author, for his technical support and practical viewpoint of the lightly loaded foundation industry. I am very appreciative of the large amount of time that Dr. Shen spent reviewing the chapters and his willingness to meet with me during weekends and even on his vacation. I am also appreciative of the contributions of Dr. Bruce Kutter of the University of California at Davis, Department of Civil Engineering.

The ADSC: The International Association of Foundation Drilling should be commended for supporting a project that benefits the entire foundation industry. Special thanks to the executive director of the ADSC, Mr. Scot Litke, as editor, for his helpful ideas, comments, time, and enthusiasm. I would also like to thank both Scot and Herb for acting as liaisons with the ADSC review committee, and Mr. Ed Nolan of Malcolm Drilling, Kent, Washington, as chairman of the ADSC Standards and Specifications Committee, for providing helpful comments and suggestions, and thank you to the rest of the members of the ADSC review committee.

Many thanks also to Mr. Ron Bajuniemi of Kaldveer and Associates, Oakland, California, and Dr. Richard Fragaszy of Washington State University, Pullman, Washington, for their in-depth review and invaluable comments. Special appreciation goes to Geocon Incorporated, San Diego, California, for providing unlimited use of their photocopying and facsimile machines and for their continuing support of higher education.

Finally, I am forever grateful to my loving wife, Teri, who diligently helped in the review process and provided a constant source of support and understanding. Although we have been married for over four years, it probably seems like only three years to Teri, because for all intents and purposes she has been husbandless since I began working on this project. I can honestly say that this guide would not exist if I had not met Teri and that I am proud to be married to such a loving, supportive, and caring person.

FOUNDATIONS IN PROBLEM SOILS

CHAPTER 1

Introduction

PURPOSE

The purpose of this guide is to expose the reader to current United States design concepts and construction practices for lightly loaded foundation systems. For our purposes, lightly loaded foundation systems are defined as foundations for one- and two-story residential, commercial, and industrial buildings. Sites that have traditionally been considered more challenging or difficult are emphasized. Examples of such sites include those with expansive soils; soft and/or organic soils; uncontrolled or deep fills; slopes; limestone solution cavities/sinkholes, and frost problems. There are two reasons for emphasizing these challenging sites. First, a large percentage of "good" sites have already been utilized. Second, upward spiralling construction costs and land values have caused owners to become more aware of the potential problems associated with structural movements, which can be magnified when building upon a difficult site. For example, it has been estimated that by the year 2000, the cost of damage to buildings constructed at sites with expansive soils will exceed $4.5 billion.[1] This estimate does not include other challenging site

[1]Chen, F.H., "Current Status on Legal Aspects of Expansive Soils." *Proceedings of the 6th International Conference on Expansive Soils.* 1987, pp. 353–356.

conditions such as those listed above. Thus, understanding the potential problems at a site and finding acceptable solutions to the problems are very important.

Designing and constructing lightly loaded foundations is not considered, in general, to be a challenging problem on good or even moderately difficult sites, relative to the design and construction of a heavily loaded foundation system. However, when constructing upon more challenging sites, the conditions encountered must be carefully considered for lightly loaded foundations. Under these conditions, the soil investigation and foundation design deserve as much attention as they would for a heavily loaded system. However, this typically is not the case for lightly loaded systems because of tight budget constraints and the desire of many developers to maximize profits.

It is not the intent of this guide to imply that the foundation industry for lightly loaded structures is in a state of chaos or stagnation. The vast majority of lightly loaded structures have been founded upon generally stable sites and have little or no foundation problems. However, those that are constructed on challenging sites are more likely to have problems.

It is the intent of this guide to examine some of the more challenging lightly loaded foundation design and construction problems. Information was gathered from specific regions in the United States with known conditions that pose challenges to lightly loaded foundations. Descriptions of these adverse conditions and the construction procedures used to alleviate them are described. It was, of course, not possible to visit or investigate every location in the country. Instead, key locations were chosen that illustrate typical problematic conditions and the methods with which these conditions are addressed. It is intended that readers in locations that have not been used as examples in this guide will be able to apply the information illustrated in the examples to their situations. Readers are cautioned, however, that each individual site is unique and should be approached as such.

It is the intent of this guide to examine some of the more challenging lightly loaded foundation design and construction problems.

This guide should not be considered a design manual. It is an informational document intended to give the reader an understanding of why different foundation systems are used, both on a regional and site-specific basis. References are given to aid in the design process after a feasible foundation system has been chosen.

GOALS

One of the goals of this guide is to present different foundation alternatives in an effort to reduce the regional nature of the industry. Traditionally, local geologic or soil conditions and the abundance or scarcity of certain building materials have played an integral part in the selection of a particular foundation system. Furthermore, the system of choice is often adopted because of a local familiarity with its design and construction, and historical precedent. Although a local system may function adequately, systems utilizing advanced technology may have advantages, such as long-term reduced costs or ease of construction, that a local industry may not be aware of. This is especially true when challenging soil or site conditions exist.

Another goal of this guide is to illustrate the importance of considering long-term costs and benefits of a proposed foundation system as well as those in the short term. Challenging sites typically increase the potential for a foundation failure to occur in the future. Therefore, it may be prudent to construct a more costly foundation system (in terms of up-front costs) to decrease the potential of incurring remedial costs in the future. An example of this may be the decision to construct a deep foundation, such as a drilled shaft or driven piling system rather than a conventional spread footing foundation. The deep foundation may cost more to construct, yet it may also provide the necessary factor of safety to ensure that there will not be significant problems during the design life of the structure.

Another goal of this guide is to illustrate the importance of considering long-term costs and benefits of a proposed foundation system as well as those in the short term. Challenging sites typically increase the potential for a foundation failure to occur in the future.

In an effort to lower the failure rate of lightly loaded foundations, an additional goal of this guide is to discuss common construction and design oversights. Conversations with engineers and contractors from around the country have indicated that a foundation failure may be defined as any movement of the foundation that is unacceptable to the owner. This is a very broad definition. However, our litigious society has created a mind-set where any defect, regardless of the effect on the performance of the building, may be considered a foundation failure in the owner's perception. If developers, contractors, engineers, and building inspectors are more aware of common oversights, the standards of construction will be further improved and the occurrence of foundation failures (or perceived foundation failures) should be reduced.

This guide also discusses simple foundation maintenance techniques, such as proper landscaping and drainage. The owner should be aware of the importance of foundation maintenance to ensure that the foundation system performs adequately for the design life of the structure.

INFORMATION SOURCES

Information in this guide has been obtained from four sources: (1) soil mechanics and foundation engineering textbooks; (2) literature prepared by professional associations with an interest in the foundation industry; (3) industry journal articles; and (4) personal interviews with engineers and contractors. A large majority of the "hands-on" information was obtained from personal interviews conducted in different geographical areas of the United States. The interviewees included architects, geotechnical engineers, contractors, insurers, and structural engineers. The geographical areas chosen represent a variety of challenging soil and site conditions as well as common local practices of some of the nation's major metropolitan regions. Readers should be aware that the interviews represent a qualitative survey of local practices and they should not be interpreted as statistically representative of the total population.

If developers, contractors, engineers, and building inspectors are more aware of common oversights, the standards of construction will be further improved and the occurrence of foundation failures (or perceived foundation failures) should be reduced.

Sources (1) through (3) provided valuable information regarding general soil mechanics and foundation engineering, specific foundation systems, or site conditions. These references are presented at the end of each chapter for readers interested in specific information regarding a particular design or site condition.

FORMAT

Chapters 2 and 3 discuss the rudiments of soil mechanics and the basic foundation types that are commonly used for lightly loaded structures. The chapter on soil mechanics explains in layman's terms the difference between clayey and non-clayey soils, the potential for settlement, the importance of proper soil compaction, and additional pertinent soil properties. The foundation types discussed include spread footings, continuous wall footings, monolithic slabs, drilled shafts, and driven piles.

Chapters 4 through 9 discuss different soil or site conditions and present applicable foundation systems and/or construction practices. Because these chapters have been written to stand on their own, the reader may notice some repetition. The foundation systems discussed in each chapter are presented from a regional perspective. An example of this is the different foundation systems that have been used for sites with highly expansive soils (see Chapter 4). In Denver, drilled shafts are commonly used; in Dallas, drilled shafts are used as well as post-tensioned or conventional heavily reinforced slabs; in San Diego, foundations on expansive soil deposits are typically either heavily reinforced continuous wall footings with floating floor slabs or post-tensioned reinforced slabs. Details of the different foundations systems are explored in these chapters as well as common design and construction oversights related to each foundation system.

Chapter 10 discusses the importance of quality control and inspection. Guidelines for inspection are presented for different foundation systems. Additionally, maintenance suggestions are presented in order to enhance the long-term performance of the foundation system. Chapter 11 examines some of the innovations that have recently entered the lightly loaded foundation industry, along with some final comments regarding the industry itself.

When consulting this guide, readers are encouraged to keep in mind the following observations:

1. Challenging soil and site conditions warrant very close attention during both the design and construction phases of a project. Experienced geotechnical consultants should be retained and their recommendations adhered to. It is always less expensive to conduct a proper geotechnical investigation up front than to perform repair procedures in the future.
2. Solutions to challenging soil and site conditions may exist outside of the reader's locale. Thus, readers are encouraged to consider solutions from other regions as alternative approaches.
3. The most cost-effective system is not always the one that is the least expensive to construct. The initial construction costs must be compared with the likelihood that foundation remediation costs will be incurred in the future.

CHAPTER 2

Soil Mechanics For Lightly Loaded Foundations

INTRODUCTION

The purpose of this chapter is to discuss the behavior and engineering properties of soil that are relevant to lightly loaded structures. These properties include the plasticity of the soil, the potential for volume change (settlement and expansion), and the moisture content. The presence of groundwater, soil compaction, and slope stability are also discussed. Common methods of testing the soil are presented. References are made in this guide to tests standardized by the American Society for Testing and Materials (ASTM). Specific tests are identified by number, for example, ASTM D-1557-78.

Soil consists of three basic constituents: solid particles, water, and air. The voids (or pore spaces) between the solid particles are filled with water and/or air. Figure 2.1 shows a typical cross section of a mixture of soil. If the void space contains only air, the soil is defined to be dry. If the void space contains only water, the soil is defined to be saturated.

Continuous mechanical weathering and chemical weathering of geologic formations create soil particles. Examples of mechanical weathering include wind and water erosion, glacial erosion, and degradation caused by freeze and thaw cycles. The result of mechanical weathering is the creation of soil grains. Chemical weathering occurs when rocks are in

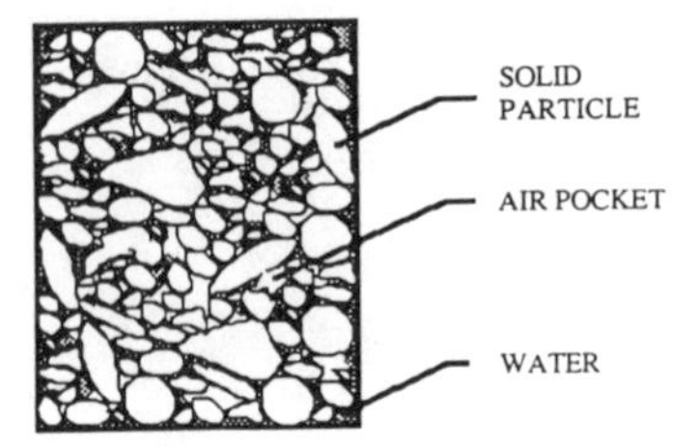

Figure 2.1 Cross section of a mass of soil.

prolonged contact with water and/or air and the chemical structure of the rock is altered. The mechanical weathering process and chemical degradation continually reduce the size and structure of the rocks and eventually the rocks become particles in the soil.

Because soil is made up of three different constituents, it is difficult to characterize the behavior of soil. The relative amounts of each constituent in a given soil deposit can vary, as can the size and mineralogy of the particles. Variations in the constituents of two soil deposits will lead to differences in their engineering properties. For example, a given volume of soil will vary in behavior depending on the predominant size of the particles. Classification systems have been adopted to help determine the predominant constituents of a soil deposit and to aid in the prediction of its engineering behavior. One of the most commonly used classification systems in the United States is the Unified Soil Classification System (USCS) originally developed by Casagrande in 1948 and modified by the U.S. Army Corps of Engineers and the U.S. Bureau of Reclamation in 1952 (Holtz and Kovacs, 1981).

Both the size of the particles and the soil's plasticity are the basis of the USCS. Soils are classified by the percentage (by weight) of different sized particles. Particle sizes are determined by sieving a representative sample of the soil through screen openings of various sizes. A soil is described as containing predominantly coarse-grained particles or containing predominantly fines. *Coarse-grained particles* are defined as either gravel or sand, depending on the particle size. *Fines* are defined as particles passing the No. 200 sieve and are either silt or clay, depending on their plasticity and Atterberg limits (Atterberg limits are discussed later in this chapter). It should be noted that clay particles are not solely defined by their size but also by their mineralogy. This aspect of clay particles is discussed in more detail in the section on non-clayey versus clayey soils later in this chapter.

	Boulders	Cobbles	Gravel	Sand	Fines (silt and clay)
Size (mm)	300	75	4.75	0.075	
U.S. Sieve No.	12″	3″	4	200	

Figure 2.2 USCS particle sizes.

The USCS classifies any particles larger than the 3-inch (76mm) sieve as "oversized" and although their presence is noted, they are not included in the initial classification of the soil. The oversized particles are defined to be cobbles if they are between 3 inches and 12 inches in diameter and are defined to be boulders if they are over 12 inches in diameter. Figure 2.2 shows the breakdown in size of the USCS classifications and the corresponding U.S. Standard sieve number.

In general, a sample collected in the field will rarely consist of only one particle size. Furthermore, the classification of an actual soil deposit is usually a combination of two or more of the soil types defined above. For example, a sample containing mostly gravel-sized grains (>50% by weight) and appreciable sand-sized grains (>12% by weight) would be defined as a sandy gravel. A sample that does not contain greater than 50% of one of the particle sizes is classified on the basis of the particle size that makes up the highest percentage, by weight, of the sample. Samples are classified as well graded or poorly graded depending on the distribution of particle sizes within a sample. Qualitatively, a soil sample having no excessive amounts of one size particle and no gaps in another size particle would be considered well graded (French, 1989). Readers are referred to soil textbooks for a quantitative discussion of poorly versus well-graded soils (French, 1989; Lambe and Whitman, 1969).

One additional soil classification is organic soil. This classification is based on the presence of any organic material in the soil and its plasticity; it is not based on any specific particle size. A soil deposit consisting mainly of organic material is defined as peat. Clays and silts can also contain organic material and are classified as organic clays and organic silts of low or high plasticity. The presence of organic material is important because organic deposits are continually decaying; consequently, erecting structures upon these deposits is not recommended.

Figure 2.3 illustrates the USCS in detail. An in-depth explanation of the USCS is beyond the scope of this manual. Readers are referred to

Major Divisions			Group Symbols	Typical Names	Classification Criteria		
Coarse-Grained Soils More than 50% retained on No. 200 sieve*	Gravels 50% or more of coarse fraction retained on No. 4 sieve	Clean Gravels	GW	Well-graded gravels and gravel-sand mixtures, little or no fines	Classification on basis of percentage of fines Less than 5% pass No. 200 sieve GW, GP, SW, SP More than 12% pass No. 200 sieve GM, GC, SM, SC 5% to 12% pass No. 200 sieve Borderline classification requiring use of dual symbols	$C_u = D_{60} D_{10}$ Greater than 4 $C_2 = \frac{(D_{30})^2}{D_{10} \times D_{60}}$ Between 1 and 3	
			GP	Poorly graded gravels and gravel-sand mixtures, little or no fines		Not meeting both criteria for GW	
		Gravels with Fines	GM	Silty gravels, gravel-sand-silt mixtures		Atterberg limits plot below "A" line or plasticity index less than 4	Atterberg limits plotting in hatched area are borderline classifications requiring use of dual symbols
			GC	Clayey gravels, gravel-sand-clay mixtures		Atterberg limits plot above "A" line and plasticity index greater than 7	
	Sands More than 50% of coarse fraction passes No. 4 sieve	Clean Sands	SW	Well-graded sands and gravelly sands, little or no fines		$C_u = D_{60}/D_{10}$ Greater than 6 $C_2 = \frac{(D_{30})^2}{D_{10} \times D_{60}}$ Between 1 and 3	
			SP	Poorly graded sands and gravelly sands, little or no fines		Not meeting both criteria for SW	
		Sands with Fines	SM	Silty sands, sand-silt mixtures		Atterberg limits plot below "A" line or plasticity index less than 4	Atterberg limits plotting in hatched area are borderline classifications requiring use of dual symbols
			SC	Clayey sands, sand-clay mixtures		Atterberg limits plot above "A" line and plasticity index greater than 7	
Fine-Grained Soils 50% or more passes No. 200 sieve*	Silts and Clays Liquid limit 50% or less		ML	Inorganic silts, very fine sands, rock flour, silty or clayey fine sands	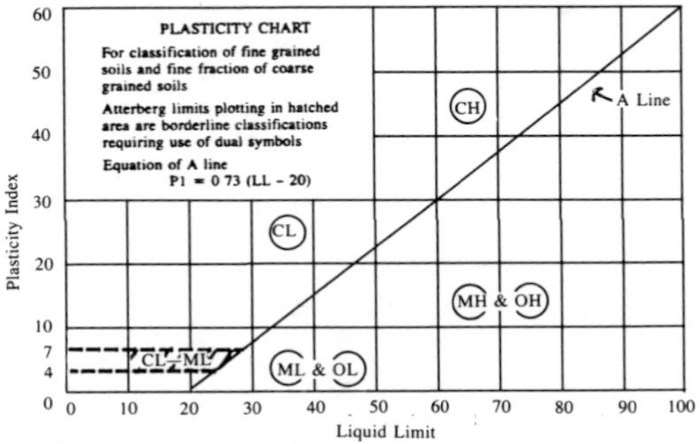 PLASTICITY CHART For classification of fine grained soils and fine fraction of coarse grained soils Atterberg limits plotting in hatched area are borderline classifications requiring use of dual symbols Equation of A line PI = 0 73 (LL - 20) Plasticity Index: 0, 4, 7, 10, 20, 30, 40, 50, 60 Liquid Limit: 0, 10, 20, 30, 40, 50, 60, 70, 80, 90, 100 CH; CL; MH & OH; ML & OL; CL-ML; A Line		
			CL	Inorganic clays of low to medium plasticity, gravelly clays, sandy clays, silty clays, lean clays			
			OL	Organic silts and organic silty clays of low plasticity			
	Silts and Clays Liquid limit greater than 50%		MH	Inorganic silts, micaceous or diatomaceous fine sands or silts, elastic silts			
			CH	Inorganic clays of high plasticity, fat clays			
			OH	Organic clays of medium to high plasticity			
Highly Organic Soils			PT	Peat, muck, and other highly organic soils	Visual-Manual Identification, see ASTM Designation D 2488.		

*Based on the material passing the 3-in. (75-mm.) sieve.

Figure 2.3 *The Unified Soil Classification System.* (Reprinted by permission of the American Society of Testing and Materials.)

any standard soil mechanics textbook for a more detailed discussion (Lambe and Whitman, 1969; Holtz and Kovacs, 1981; Casagrande, 1948; Howard, 1986).

NON-CLAY VS. CLAY SOILS

Non-clay (granular) soils consist predominantly of gravel, sand, and silt-sized particles. Interaction between the particles consists only of frictional forces, which act parallel to the plane of contact. The frictional forces keep the particles from sliding past one another. Because of this, there is a limit to the steepness an unconfined surface of dry, non-clay soil can approach before gravitational forces overcome the frictional forces. The addition of some water to the soil may increase the potential for the surface of the soil to approach vertical. This is due to surface tension among the air, water, and particles, which helps hold the particles together. The magnitude of the surface tension increases as the size of the pore spaces decreases. Thus the surface tension developed in a gravelly soil will be relatively small and is typically not large enough to enable a temporary vertical face to be constructed. However, the vertical face of a sandcastle built with most silty sand illustrates the effects of surface tension on soils with small particles. If the moisture in the sandcastle evaporates, it will collapse. The rate that moisture within the soil will flow to the surface and evaporate is dependent on the permeability of the soil and the magnitude of the surface tension. The sandcastle will dry fairly rapidly because the surface tension between the sand particles is relatively small. Therefore the moisture within the sandcastle is not readily drawn up to the drying surface. A clayey soil is able to maintain a vertical face longer than a sandy soil because the surface tension forces are large compared to the size of the particles. The low permeability of the clay enables it to retain its moisture longer.

The phenomenon of the collapse of temporary, unconfined vertical surfaces comprised of granular soils is important for lightly loaded construction because it may make completing footing excavations and/or drilled shafts difficult owing to sloughing or collapsing of the sides.

Clay minerals are flat and platy in shape due to their molecular structure. The platy shape provides a broad surface for electrochemical interaction to occur between the particles. Clay minerals also have an overall negative electrical charge. The platy shape and electrical charge enable the clay minerals to interact both chemically and electrically with the pore fluid, which typically has positive ions in suspension. The interaction with the pore fluid creates a thin layer around each of the clay particles. This thin layer of water is known as adsorbed water (Holtz and Kovacs, 1981). The chemical interaction and electrical forces between the particles and the interaction with the pore fluid are strong enough to keep them together. Common clay minerals include kaolinite, illite, and montmorillonite. An excellent discussion of the mineralogy and structure of clay minerals is provided by Holtz and Kovacs (1981).

Clayey soil deposits typically have very low permeabilities. The small size of the clay particles, the flat and platy shape of the particles, and the affinity for adsorbing water inhibit the flow of water through the soil. The presence of moisture in a clayey soil keeps a temporary vertical cut from collapsing. However, as discussed in later sections of this chapter, if a clayey deposit has a very high water content its strength will decrease and it will not be stable in a vertical (or near vertical) cut. Additionally, a vertical surface of a clayey soil will eventually collapse, as the granular soil will, when it completely dries out. The main difference is that the time that it will take for the clayey soil to dry is much longer.

Silty soils are fine-grained soils such as clayey soils, yet they are typically not as plastic (see section on Atterberg Limits). The permeability of a silty soil is low, but not as low as a clayey deposit. The silt particles do not tend to adsorb water as clay particles do.

Sand and gravel deposits are described as very loose to very dense. There are varying degrees between these extremes that depend on the soil's density, or state of packing, in the field. Fine-grained soils, such as clayey and silty deposits, are described as very soft to very hard with varying degrees in-between, depending on their consistency in the field. Soil deposits have varying in situ densities and consistencies, depending on their original depositional environment, as well as the amount of loading they have undergone since their deposition. The consistency or density of the soil is typically determined in the field by conducting a Standard Penetration Test (ASTM D-1586-67). This test is performed by counting the number of blows it takes a 140-pound weight, dropping 30 inches per blow, to drive a 2-inch outside diameter sampling barrel 18 inches into the soil. The number of blows it takes to drive the barrel the final 12 inches is recorded as the standard penetration number.

The presence of clayey and silty soils enables excavations and drill holes to remain open more readily. However, these soil deposits are more likely to present other problems such as settlement, swelling, low shear strength, and frost heave.

MOISTURE CONTENT AND THE GROUNDWATER TABLE

The amount of water in the soil has a large effect on many engineering properties of a soil deposit. The presence of water will affect clayey soils differently from non-clayey soils and will generally have a more adverse effect on clayey soils. The moisture content is defined as the ratio, by weight, of water to solids in the soil. It is not defined as the percentage of the voids that are filled with water. The latter is defined as the degree of saturation. Thus, a saturated condition does not imply that the moisture content is 100%. For very loosely packed saturated soils, moisture contents typically will be in the range of 40–60%.

The microscopic interaction between water and clay minerals was discussed in the previous section. Other aspects of the behavior of clayey soil that are affected by moisture are compaction, consolidation, loss of strength, volume change, and slope stability. These properties will be discussed in the following sections of this chapter.

The behavior of non-clayey soils affected by moisture includes compaction, slope stability, sloughing, and frost heave. These effects are discussed in the subsequent sections and chapters.

Groundwater can also create difficulties during construction. Examples of these difficulties include its effect on earthwork, basement construction below the water table, footing excavations into the water table, and the placement of concrete through the water table. These difficulties will be discussed in later chapters.

ATTERBERG LIMITS

In 1911 a Swedish scientist named Atterberg recognized that clayey soils exhibit markedly different consistencies depending on their moisture content. It was observed that these consistencies ranged from that of a brittle solid to a liquid. Atterberg defined dividing lines, such as the plastic limit and the liquid limit, between the different consistencies (ASTM

4318). The plastic limit (PL) is the moisture content below which a soil no longer behaves in a plastic manner. The liquid limit (LL) is defined as the moisture content above which a soil no longer behaves plastically and has the consistency of a very viscous fluid. The method of determining the plastic and liquid limits is outlined in any standard soil mechanics textbook (Atterberg, 1911; Lambe and Whitman, 1969; French, 1989; Holtz and Kovacs, 1981).

The plasticity index (PI) (ASTM 4318) is defined as the difference between the liquid limit and the plastic limit. The plasticity index has been used empirically to predict different engineering properties. An example of this is the empirical relationship between the PI and swell potential (Chen, 1988; Holtz and Gibbs, 1954; Seed, Woodward, and Lundgren, 1962).

EXPANSION AND SHRINKAGE POTENTIAL OF CLAYS

As will be discussed further in Chapter 4, expansive clays present some of the most challenging soil conditions upon which lightly loaded structures are constructed. Some clays can impart large uplift pressures when they expand and lightly loaded structures do not have enough weight to counteract these large uplift forces.

> Expansive clays present some of the most challenging soil conditions upon which lightly loaded structures are constructed.

As mentioned previously, clay particles interact with water. Certain clay minerals are able to "take on water" and subsequently expand when they come in contact with it. It is beyond the scope of this book to explain in detail the mechanism of expansion of clayey soils, and many aspects of it are not fully understood. A thorough discussion can be found in Chen (1988) and Holtz and Kovacs (1981). Figure 2.4 shows regions in the United States that have extensive deposits of soils with high to moderate expansion potentials.

As the source of water is removed and a clayey soil begins to dry out, it shrinks. In general, the shrinkage phenomenon is the reverse of the expansion process. As the water evaporates, surface tension between the

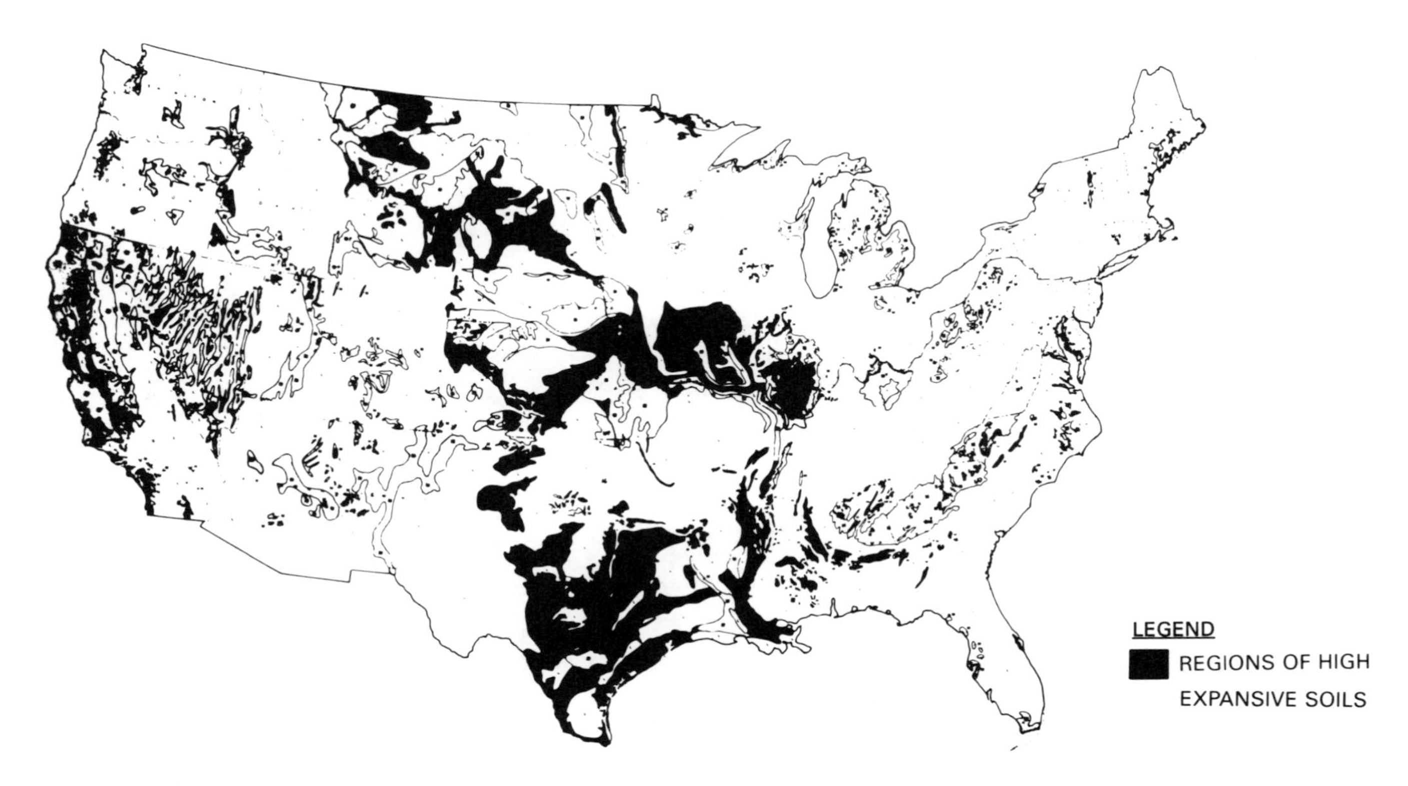

Figure 2.4 *Expansive soils in the United States.* (Adapted from Wiggins, 1976, and reprinted by permission of the Post-Tensioning Institute.)

water and the particles causes the particles to be pulled together. Soils with high expansion potential also tend to have high shrinkage potential. Foundation alternatives for sites with expansive soils are discussed in Chapter 4.

SETTLEMENT: POTENTIAL AND MECHANISM

There are two types of settlement resulting from an increase in loading on a soil mass: immediate settlement and consolidation settlement. Immediate settlement occurs as soon as a load is applied to the soil. It is caused by the deformation of the particles in the soil due to the extra weight. In essence, the individual soil particles are being "squeezed," similar to what would happen if a heavy weight were placed on top of a rubber eraser. According to Sowers (1979), immediate settlement for clayey soils is seldom critical because "a clay strong enough to support a footing will be rigid enough that the distortion [immediate] settlement usually will not be serious. It occurs rapidly, during construction, and usually goes unnoticed." It is also noted that a future increase in building loads due to a new addition, for example, may result in additional immediate settlements.

Consolidation settlements occur in saturated or near-saturated soil deposits. When a saturated or nearly saturated soil is loaded, the pore fluid pressure increases. This increase is defined as excess pore pressure. The excess pore pressure causes the pore fluid to be expelled from the soil. As the pore fluid is forced from the soil, the increased load is transferred to the soil particles. The increased load on the soil particles compresses them together, thus consolidating the soil.

The amount of consolidation depends on the compressibility of the soil structure and the magnitude of the applied load. Compressibility of the soil structure may be likened to a spring. When a spring is subjected to a load, it compresses. There are different compressibilities of different soil deposits just as there are different compressibilities of different springs.

Consolidation settlements are time-dependent. They occur in both clayey and non-clayey soils. However, the rate of the consolidation settlement is dependent on the permeability of the soil. The permeability of non-clayey soils is relatively high and therefore the consolidation settlement occurs rapidly and is usually assumed to occur instantaneously.

Clayey soils have very small pore spaces—therefore, it takes time for the excess pore pressure to dissipate as the water is extruded from the soil. The consolidation process never truly stops; for nonorganic soils, however, the settlement will eventually become insignificant from an engineering standpoint. The rate of consolidation is dependent on the distance water has to travel to be expelled from the clay and the permeability of the soil. It can take up to several years for the majority of the consolidation to occur if the clayey deposit is relatively thick and/or has a fairly low permeability.

Soils that have an organic constituent have a large potential for settlement because they are continually decaying. Construction of a foundation that will impart any additional load on organic soils, regardless of how light, is not recommended. Some examples of construction techniques for sites with organic soil deposits are discussed in Chapter 5.

EXPANSION OF SILTY SOILS DUE TO FROST

In many parts of the country the temperature becomes cold enough during the winter to freeze the water in the soil. The depth at which the water freezes depends on the duration of the subfreezing temperatures. Local conditions should be taken into account when determining the potential frost depth. Local building departments typically specify the design depth of frost penetration. When the soil freezes, it can increase in volume by as much as 10%. This is due to the increase in volume of the water in the voids when it becomes ice (Holtz and Kovacs, 1981) and to the continued growth of ice lenses in the soil.

If there is a relatively shallow groundwater table beneath the frozen ground, capillary action in the soil can continue to feed the frozen area with water. Capillary action is the phenomenon of drawing moisture up into the soil above the groundwater table, similar to a paper towel soaking up a spill. The distance the moisture can rise is dependent upon the average pore space. The smaller the pore space, the higher the rise.

As the capillary action continues to feed the frozen area of soil, an ice lense forms and grows. This ice lense is strong enough to push up the soil and any lightly loaded structures on the surface. When spring arrives, the ice melts and the potential for settlement increases.

Silty soils are more susceptible to frost heave than are other soil types. The relatively high capillary rise of silty soils increases the potential for a continual source of moisture to feed the growing ice lenses. Clay soils also have a high capillary rise, but low permeabilities inhibit rapid moisture flow during the cold period. Frost heave and other cold weather foundation difficulties are discussed in Chapter 7.

COMPACTION

Compaction is defined as the process of forcing air out of the soil, resulting in an increase in density. This is commonly achieved in the field by either (1) rolling or kneading the soil, (2) vibrating the soil, or (3) dropping a heavy weight on the soil. The first method is most effective for clayey soils, the second method is most effective for non-clayey soils, and the third method is effective for both types of soils.

Compaction in the field is controlled by determining the dry density of the soil as compacted and comparing it to the maximum dry density of the same soil as determined by a laboratory test. Two commonly used laboratory tests determine the maximum dry density and the corresponding optimum moisture content of a field sample of soil: the standard Proctor (ASTM D-698) and the modified Proctor (ASTM D-1557) tests. These laboratory tests consist of placing soil in layers (lifts) into a standard mold and compacting them by dropping a standardized hammer a specified number of times. Once the mold is full, it is weighed and the density of the material can be determined. A portion of the soil is oven-dried to determine the moisture content by weight. The test is conducted at varying moisture contents in order to obtain a range of points along the compaction curve (Figure 2.5) for the soil in question. The basic difference between the standard and the modified Proctor is the amount of energy (the weight of the hammer and the height from which it is dropped) that is forced into the soil and how many lifts in which the soil is compacted during the test. The modified Proctor energy is approximately 4.5 times greater than the energy of the standard Proctor test. The standard test was developed in the early 1930s to monitor the compaction of pavement subgrades. The modified test was established to better simulate the modern equipment being used to construct airfields during World War II (Holtz and Kovacs, 1981), and building codes have since accepted the construction of foundations on engineered fills using the modern equipment and subsequently higher soil densities.

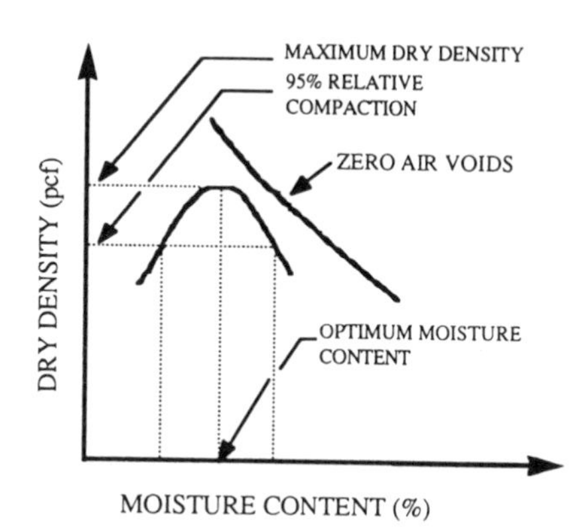

Figure 2.5 Typical compaction curve for a given compactive effort.

The amount of compaction that can be achieved for a given soil is dependent on (1) the compactive effort and (2) the water content of the soil. The compactive effort refers to how much soil is pounded, vibrated, or kneaded. Figure 2.5 shows a typical laboratory compaction curve of dry density versus water content for a specific compactive effort. The maximum dry density is shown at the peak of the curve. The water content at which the maximum dry density is achieved is defined as the optimum water content. As can be seen in Figure 2.5, a specific relative compaction (for example, 95% as shown in the figure) can be achieved for this soil over a range of moisture contents. Moisture contents below the optimum moisture content are defined to be on the "dry" side of optimum and, conversely, moisture contents above the optimum moisture content are on the "wet" side of optimum. The properties of a compacted soil vary depending upon whether they have been compacted on the dry or wet side of the optimum moisture content (Seed, 1959).

In general, a clayey soil compacted on the dry side will have a higher strength and less potential for shrinkage than one compacted on the wet side. However, when compacted on the wet side, the soil will have a lower permeability and a lower potential for expansion (Seed, 1959). The method of compaction will also affect the properties of the compacted clayey soil. Different methods include static compaction, such as a roller; kneading compaction, such as a sheepsfoot compactor; and vibratory compaction, such as a vibratory roller. In general, the method of compaction has a significant effect on the compacted clayey soil when the compaction is conducted on the wet side of optimum. The strength of the compacted soil will be highest if a kneading method is used, less so if a vibratory method is employed, and even less if a static method is used. Similarly, the shrinkage potential is highest for kneading methods, less for vibratory methods, and even less for static methods. Therefore, it is important to evaluate the intended performance of the compacted clayey soil and compact it on the appropriate side of the optimum water content.

When a soil sample is compacted on the dry side of the optimum water content, the friction between the particles limits the density of the sample that can ultimately be achieved. As water is added to the soil it provides lubrication between the particles and a higher density can be attained. As more water is added to the soil it becomes more difficult for the air in the voids to escape and the air is sealed in. The addition of water to the void space without an increase in the packing of particles leads to a decrease in the density of the soil, because the water and the "trapped" air have replaced the solid particles in a given volume of soil. This is true for all soil types except "clean" (lacking fines) free-draining sands and gravels. The water drains so rapidly from these soil types that they typically do not have an optimum water content. An efficient method of compacting these types of soil is to puddle or flood them (Monahan, 1986).

Once a laboratory compaction curve is established, the dry density of the soil is determined in the field and the amount of compaction achieved is calculated as a percentage of the maximum dry density that was determined in the laboratory using a *specific* compactive effort (e.g., the standard or modified Proctor tests). In this manner, project compaction specifications can be monitored and additional field compactive effort applied as necessary.

There are three common misconceptions about compaction. The first is that 100% relative compaction implies that all of the air has left the soil (zero air voids). As can be seen in Figure 2.5 and Figure 2.6, the compaction curve never reaches the zero air voids curve. It is impossible to remove all of the air voids with conventional field compaction techniques. One hundred percent relative compaction means that the dry density of the soil in the field is equal to the maximum dry density achieved in the laboratory at a *specific* compactive effort.

The second misconception is also related to the concept of 100% relative compaction. The modified Proctor test has been standardized to approximate typical compactive efforts achieved in the field using modern, conventional equipment. Theoretically, it is possible to achieve relative compactions above 100%. However, because the modified Proctor energy is similar to the energy that is applied in the field with a reasonable amount of effort, it is not practical to attempt to achieve (or to

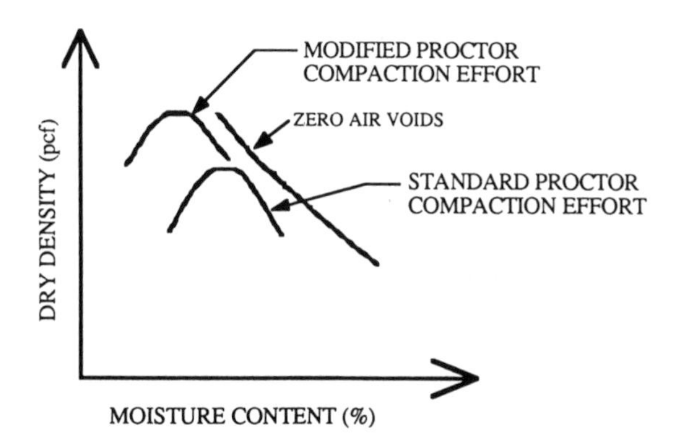

Figure 2.6 Typical standard and modified Proctor compaction curves for an identical soil sample.

specify) relative compactions at or above 100%, as determined by the modified Proctor test.

However, as illustrated in Figure 2.6, the maximum dry density of a soil sample, as determined by the standard Proctor test, is significantly lower than that as determined by the modified Proctor test. Therefore, if the field densities are compared to the standard Proctor maximum dry density, relative compaction percentages above 100% could be achieved with a reasonable amount of effort. The important concept to remember is that the relative compaction is related to a specific compactive effort and one should always determine the energy (or test) that was used to determine the "maximum" dry density.

The third misconception is that a given soil has a specific optimum moisture content. The optimum moisture content is dependent on the amount of compactive energy and, in the field, the method of compaction (Monahan, 1986). As the compactive energy increases, the optimum moisture content decreases.

The important concept to remember is that the relative compaction is related to a specific compactive effort and one should always determine the energy (or test) that was used to determine the "maximum" dry density.

From discussions with engineers and contractors around the country, it was determined that oversights related to compaction are very common reasons for problems associated with lightly loaded structures. These problems are usually associated with differential settlement of the structure. Thus an increase in soil density is desired in order to reduce the settlement potential and increase the strength. Field examples and construction techniques concerning compaction are presented in Chapters 4 and 6 and some common oversights are discussed.

SLOPE STABILITY

Slopes and hills have an inherent stability problem. The driving force affecting the stability is gravity. Gravity is constantly "pulling" on the slope in an attempt to force it into a more stable configuration. The stability of a slope depends on the soil type and strength, the presence of the groundwater table, the presence of a weak subsurface layer, and the geometry of the slope. Slopes can collapse due to a variety of reasons including: (1) their own weight, (2) the influx of water reducing the shear resistance of the soil, (3) dynamic loading from an earthquake, (4) erosive action, (5) a change of the geometry of the slope, and (6) undercutting. A phenomenon known as *creep* can also affect slope stability. Creep refers to a slow movement of the soil downslope.

The construction of a structure on or near the top of a slope will increase the likelihood of these types of failures. This increase in likelihood is due to the additional weight that the slope must now support. Additionally, irrigation of the landscaping around the new structure, leakage from swimming pools, leakage from water pipes or any other sources of water will decrease the stability of the slope as the addition of water to the soil reduces the shear resistance of the soil. Construction techniques on or near hillsides are discussed in Chapter 9.

BEARING CAPACITY

Bearing capacity is the amount of loading the soil can support before it fails or "punches" in, and is expressed in units of force per unit area. Readers familiar with soil mechanics may wonder why this important engineering property was not discussed until the end of this chapter. The reason is that in general, for lightly loaded foundation design, settlement potential controls the design. The soil may undergo significant settlement due to increased loading, yet not catastrophically fail as in a typical bearing capacity failure. Geotechnical engineers typically provide recommendations based on the local soil conditions for design-bearing capacities. However, local building specifications typically include minimum foundation dimensions. These minimum dimensions are typically large enough to reduce the contact pressure of the foundation below the bearing capacity of the soil because of relatively light column and wall loads. In cases where the bearing capacity of the soil is extremely low, potential settlements are typically extremely high and thus settlement is still the dominant factor of the design.

SUMMARY

- Soil consists of three constituents: water, air, and solid particles.
- Soils are either predominantly coarse-grained or consist mostly of fines, and they are typically classified by their dominant particle size and their plasticity.
- In general, granular soils provide better foundation support than do clayey soils (provided they are in a relatively dense state) because they are not as sensitive to moisture variations.
- The expansion, shrinkage, compaction, strength, consolidation settlement, and slope stability of clayey soils are very sensitive to changes in moisture content.
- Silty soils are the most susceptible to frost heave.
- Granular soils can create some difficulties in construction because they may slough during excavation or drilling.
- Settlement, not bearing capacity, is usually the controlling design parameter for lightly loaded foundations.
- Expansive soil deposits, deposits with high settlement potential, high water levels beneath a foundation, nonengineered fills, thick compacted fills, and building sites near or on slopes are examples of

the more common challenging site conditions that affect lightly loaded construction.

RELATED REFERENCES

ATTERBERG, A., "Lerornas Forhallande till Vatten, deras Plasticitetsgranser och Plasticitetsgrader" (The Behavior of Clays with Water, Their Limits of Plasticity and Their Degrees of Plasticity), *Kungliga Lantbruksakademiens Handlingar och Tidskrift,* Vol. 50, No. 2, pp. 132–158, 1911.

BOWLES, J. E., *Engineering Properties of Soils and Their Measurement,* 3rd Edition, McGraw-Hill Book Co., New York, 1986.

CASAGRANDE, A., "Classification and Identification of Soils," *Transactions,* ASCE, Vol. 113, pp. 901–930, 1948.

CHEN, F. H., *Foundations on Expansive Soils,* Elsevier Scientific Publishing Co., New York, 1988.

COLE, K. W., *Foundations,* Thomas Telford Ltd., London, 1988.

FRENCH, S. E. *Introduction to Soil Mechanics and Shallow Foundations Design,* Prentice-Hall, Englewood Cliffs, NJ, 1989.

HOLTZ, W. G., AND GIBBS, H. J., "Engineering Properties of Expansive Clays," Proceedings, ASCE, Vol. 80, 1954.

HOLTZ, R. D., AND KOVACS, W. D., *An Introduction to Geotechnical Engineering,* Prentice-Hall, Englewood Cliffs, NJ, 1981.

HOWARD, AMSTER K., *Visual Classification of Soils: Unified Soil Classification System,* 2nd Edition, Denver: Geotechnical Branch, Division of Research and Lab Services, Engineering and Research Center, Bureau of Reclamation, 1986.

LAMBE, T. W., AND WHITMAN, R. V., *Soil Mechanics,* John Wiley & Sons, New York, 1969.

MONAHAN, E. J., *Construction Of and On Compacted Fills,* John Wiley & Sons, New York, 1986.

POST-TENSIONING INSTITUTE, *Design and Construction of Post-Tensioned Slabs-on-Ground,* 1st Edition, Post-Tensioning Institute, Phoenix, AZ, 1989.

SEED, H. B., "A Modern Approach to Soil Compaction," *Proceedings of the California Street & Highway Conference,* 11th Proceedings, 1959, pp. 77–93.

SEED, H. B., WOODWARD, R. J. AND LUNDGREN, R., "Prediction of Swelling Potential for Compacted Clays," *Journal ASCE, Soil Mechanics and Foundations Division,* Vol. 88, 1962.

SMITH, G. N., *Elements of Soil Mechanics,* 6th Edition, BSP Professional Books, Oxford, England, 1990.

SOWERS, G. F., *Introductory Soil Mechanics and Foundations; Geotechnical Engineering,* 4th Edition, Macmillan Publishing Co., New York, p. 464, 1979.

TERZAGHI, K., *Theoretical Soil Mechanics,* John Wiley & Sons, New York, 1943.

U.S. ARMY ENGINEER WATERWAYS EXPERIMENT STATION, "The Unified Soil Classification System," Technical Memorandum No. 3-357. Appendix A, *Characteristics of Soil Groups Pertaining to Embankments and Foundations,* 1960.

WIGGINS, JOHN H., *Natural Hazards, An Unexpected Building Loss Assessment,* Technical Report NO. 1246, J. H. Wiggins Co., Redondo Beach, CA., December, 1976, pp. 95–134.

CHAPTER 3

Common Foundation Systems

INTRODUCTION

The purpose of a foundation is to transfer the weight of a structure to the soil in a manner that will not cause excessive distress to the soil or the structure. The purpose of this chapter is to describe the commonly used foundation systems for lightly loaded structures that effectively accomplish this load transfer. The foundation systems that will be discussed in this chapter are: (1) isolated spread footings, (2) continuous wall footings, (3) monolithic slabs, (4) drilled shaft[1] and grade beam systems, and (5) driven piles. These are the five basic types of foundations used in the lightly loaded foundation industry today. Many variations or combinations of these foundations have been and are presently being used. Some innovative systems have also been introduced in recent years; however, they are typically variations of or additions to the five systems referenced above. These innovations are discussed in Chapter 11.

The following is an introduction to the five foundation systems that will be discussed in more detail in subsequent chapters. A section is included at the end of the chapter that briefly discusses common floor systems constructed in conjunction with the foundations. As discussed in

[1]Also referred to as drilled piers, bored piles, and cast-in-place uncased piles.

Chapter 1, this book is a foundation guide, not a design manual. References are supplied in this chapter to provide the reader with additional information that is useful in the design process.

ISOLATED SPREAD FOOTINGS

The isolated spread footing foundation consists of square, rectangular, or circular concrete pads beneath each of the load-bearing columns of the building. The footings are constructed by excavating or drilling a hole into the soil, constructing concrete forms, placing any necessary steel, and then pouring the concrete. The floor slab is poured after the forms are stripped from the footings. The footing increases the contact area between the soil and the column, thus reducing the stress transferred to the soil. For example, a 1-foot square column may be designed to carry 10,000 lbs. If it were placed directly upon the soil the contact pressure would be 10,000 pounds per square foot (psf). However, if it is placed on a 2-foot square isolated footing, the contact pressure is reduced to 2500 psf, thus reducing the potential for a bearing capacity failure and the settlement that can be expected.

The minimum depth of the footing is usually determined by local building codes. The building department determines the minimum depth by estimating potential for frost, water erosion, and wind erosion penetration (French, 1989). The minimum depth is also dependent on the expansiveness of the soil (see Chapter 4). Figure 3.1 shows a typical

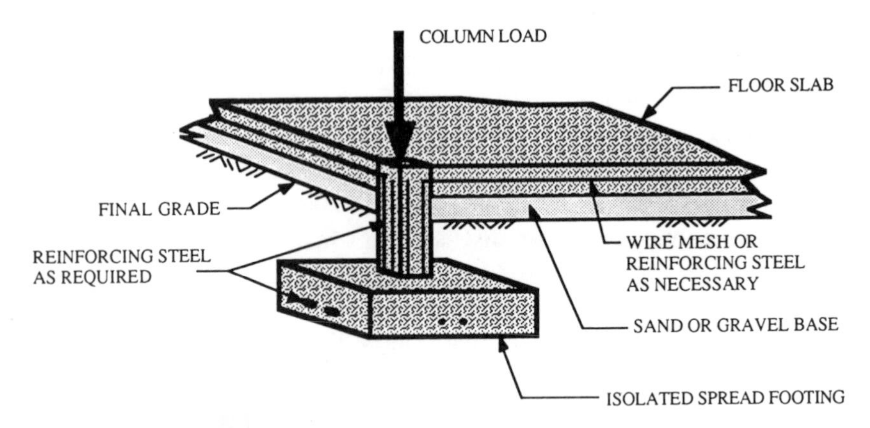

Figure 3.1 Isolated spread footing and slab-on-grade.

isolated footing design. French (1989) is a good reference concerning the design and construction of spread footings.

CONTINUOUS WALL FOOTINGS

This foundation system is used to support a load-bearing wall or a series of closely spaced column loads. Examples of load-bearing walls are a basement wall, the wall of a concrete tilt-up panel structure, or the wall of a residential structure. The continuous wall footing is constructed similarly to the spread footings, although some designs and/or soil conditions do not require any concrete forms. Instead, a uniform-width trench is excavated and the concrete is poured directly into the trench to create the footing.

Figure 3.2 shows a typical continuous wall footing design. The continuous wall footing reduces the contact pressure in the same manner as an isolated spread footing. An 8-inch-thick wall may weigh 4000 pounds per linear foot (plf) and if placed directly on the soil would create a contact pressure of 6000 psf. If the continuous wall footing is designed to be 16 inches wide, the contact pressure is reduced to 3000 psf.

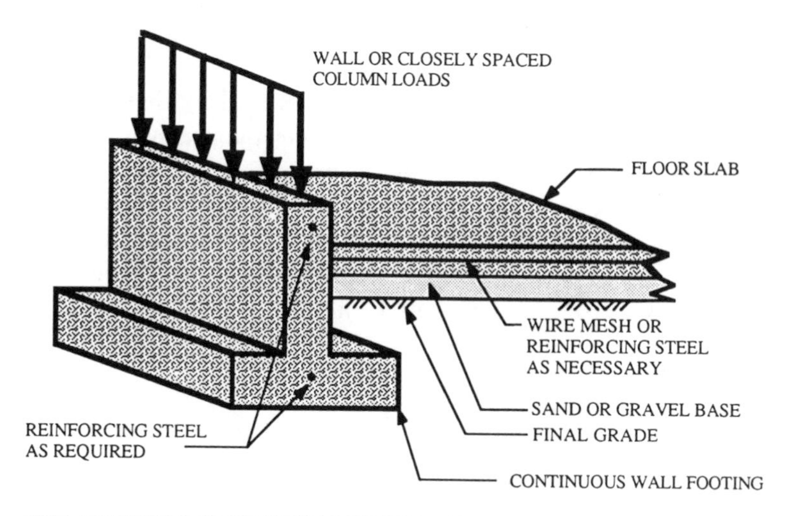

NOTE: CONTINUOUS FOOTING TYPICALLY WILL BE CONSTRUCTED AROUND THE ENTIRE PERIMETER OF THE BUILDING. THE FRONT HAS BEEN LEFT BLANK HERE IN ORDER TO SHOW THE FLOOR SLAB.

Figure 3.2 Continuous wall footing.

The minimum depth of the footing is determined by local building codes (as with isolated spread footings). Maximum allowable bearing capacities are also generally dictated by local building codes. The continuous wall footing is the most common type of foundation system for lightly loaded commercial structures and is also commonly used for residential construction. A combination of continuous wall footings around the perimeter of the building and isolated spread footings supporting interior columns is also very common as are buildings with an interconnected continuous footing grid to support interior-bearing walls.

MONOLITHIC SLABS

This type of foundation consists of a thickened edge, monolithically poured slab. The slab may also contain a series of thickened beams running in both directions in the interior. This type of design is commonly referred to as a "waffle" slab. Reinforcement of the slab can range from no reinforcement to wire mesh, conventional reinforcing bars, or post-tensioned tendons. The amount of reinforcement is usually dependent upon the soil conditions. These conditions are discussed in subsequent chapters, and design recommendations are presented in the *Uniform Building Code,* UBC Standard 29-4 (International Conference of Building Officials, 1988).

The thickened edge acts as a continuous wall footing. It is thickened to provide strength against bending under the perimeter wall load. The additional depth also helps to guard against frost, erosion, and moisture from the surface. Figure 3.3 shows a typical slab foundation with a thickened edge and thickened interior beams. The interior beams make the slab more rigid so as to keep it from bending in the event that the foundation is subjected to differential soil movement.

DRILLED SHAFT AND GRADE BEAM SYSTEM

This system consists of cast-in-place concrete shafts, with beams, commonly referred to as *grade beams,* spanning between them to support the structure (and sometimes the floor system). The shafts are constructed by drilling a hole into the ground and placing concrete into the shaft (along

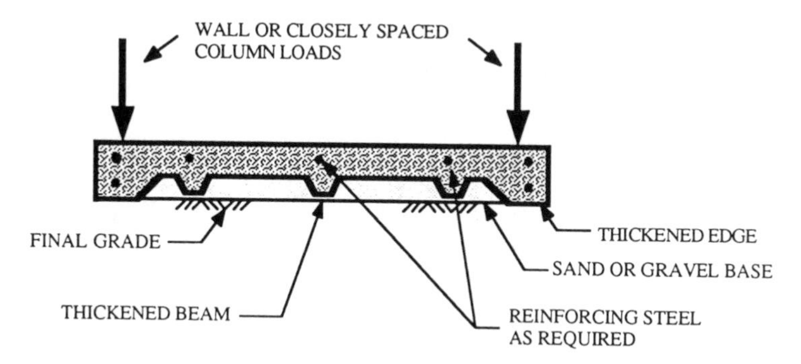

Figure 3.3 Typical monolithic slab with "waffle"-type beams.

with any necessary reinforcement). The grade beams are placed either below or above grade depending on the soil conditions. The placement of the grade beam will be discussed in subsequent chapters. Figure 3.4 shows a typical drilled shaft and grade beam system with a structural floor system and end bearing shafts.

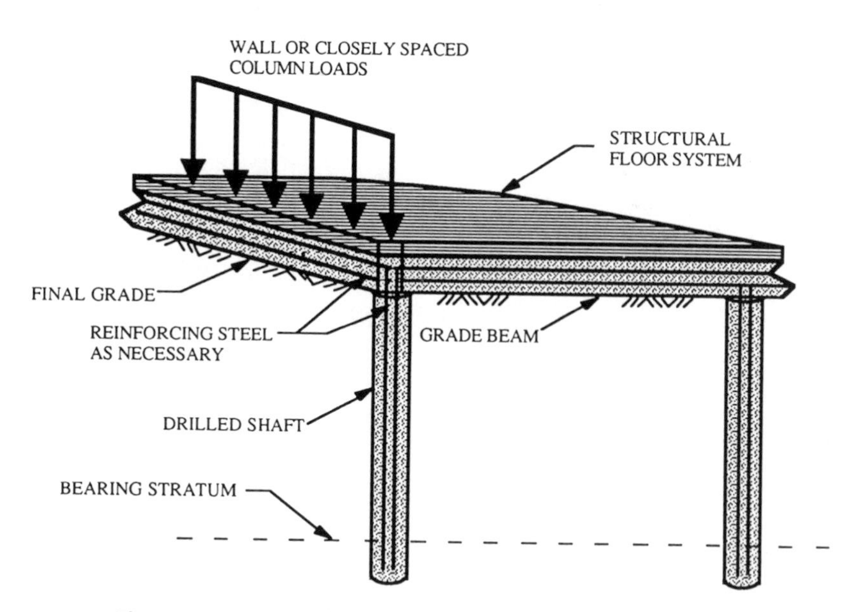

Figure 3.4 Typical drilled shaft and grade beam system.

The drilled shaft and grade beam system is a common solution for more challenging sites that do not provide adequate bearing strength near the surface. There are two different methods by which a drilled shaft can transfer the structure load to the soil. One method is the mobilization of side friction. Side friction support occurs as the shaft is loaded, mobilizing friction forces between the shaft and the soil. The structure is thus supported by the friction built up along the sides of the shaft. Friction shafts typically are straight, as shown in Figure 3.5.

Another method of load transfer to the soil is an end bearing shaft. End bearing shafts transfer the load to a soil stratum at depth, which has a higher bearing capacity than the stratum above. There are two types of end bearing shafts: straight and belled. Figure 3.5 illustrates the difference between the straight and belled shaft. Straight shafts are used in end bearing if the bearing stratum has an adequate bearing capacity to support the pressure applied to it through the base of the foundation. The

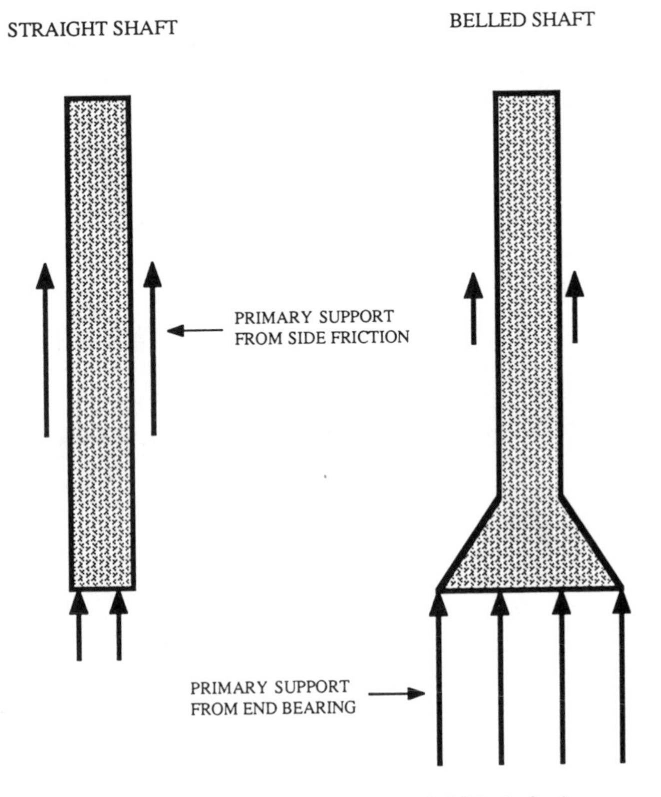

Figure 3.5 Typical variations of drilled shafts.

bell at the base of the shaft spreads out the load and reduces the contact pressure with the soil, if necessary.

As we will see in Chapter 4, drilled shafts are also used for construction in areas with expansive soils. Briefly, the shafts are used to hold the structure down by socketing them into a nonexpansive stratum.

For further design and construction methodology of drilled shafts, refer to Reese and O'Neill (1988).

DRIVEN PILES

Driven piles consist of prefabricated, relatively long structural members that are driven deeply into the ground so as to support the structure. The piles are made of timber, concrete, or steel. Steel piles are rarely used for lightly loaded structures, however; this is because they have a higher material cost, and the higher strength pile is rarely needed.

The piles are typically used in groups. Piles are driven into the soil using a pile-driving rig consisting of a crane with a hammer and leads to guide the pile. Once they have been driven into place, a concrete pile cap is constructed on top of the group of piles. At this point, work can continue as an isolated spread footing foundation or grade beams can be constructed spanning between the pile caps. Figure 3.6 shows a typical pile foundation system, in this case using tapered timber piles and a grade beam spanning between them. Another method of using piles is to install a group of them relatively close together and then construct a concrete pile cap on top of them. The pile cap then supports a column of the building. The taper of the piles increases the load-carrying capacity of the piles. Olson and Long (1989) argue that the side capacity in any given soil layer increases as the pile tip passes through that layer due to the increase in lateral strain from the increased diameter of the pile.

The two methods of transferring the load of a structure to the soil using pile foundations are similar to that of drilled shaft foundations. Friction piles transfer the load to the soil from the mobilization of side friction between the pile and the soil. Point-bearing piles transfer the structure load to a high bearing stratum of soil at the tip of the pile.

Driven pile foundations are not as commonly used for lightly loaded construction as other foundation systems owing to their relatively high installation cost. However, for some soil and site conditions, the driven pile foundation is utilized in some areas of the country. Examples

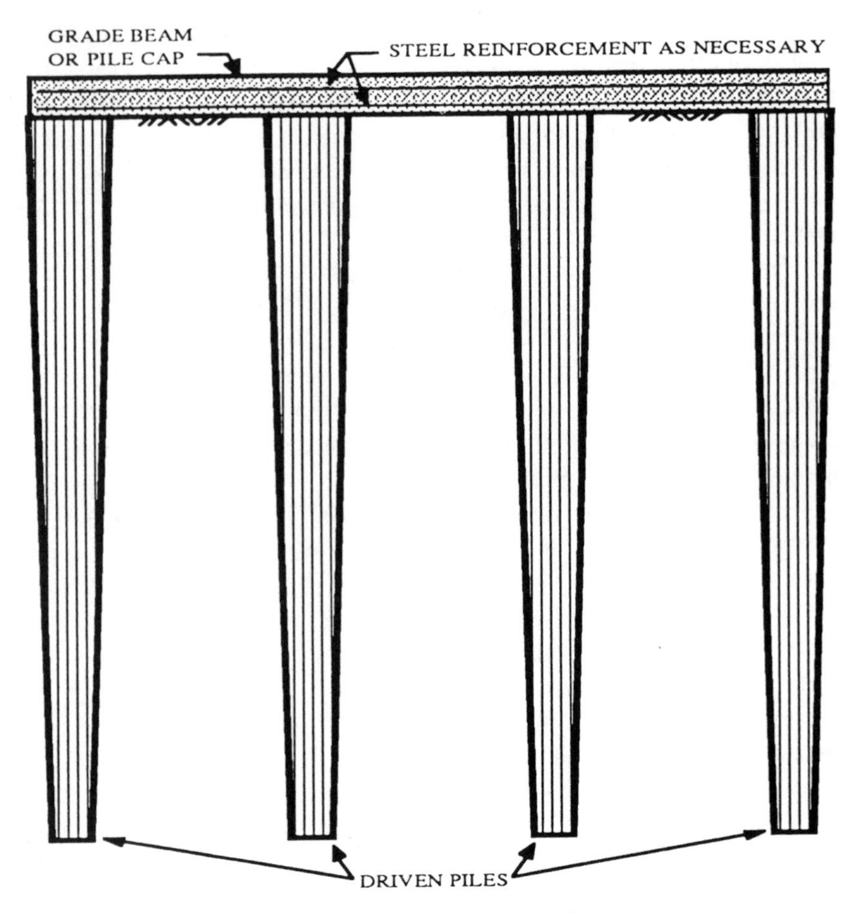

Figure 3.6 Typical driven pile foundation system.

of their use for lightly loaded structures will be discussed in subsequent chapters. For more information on the design of pile foundations refer to Fleming, Weltman, Randolph, et al. (1985) or Tomlinson (1987).

FLOOR SYSTEMS

The most common floor system typically consists of concrete poured onto a thin layer of sand or gravel that has been placed on the subgrade. Although a floor slab of a structure typically does not transfer the building load to the soil, It does frequently interact with the foundation system. Because it is commonly constructed directly on the ground surface, the floor slab design and construction should also be carefully considered, especially when a challenging building site is under consideration. For example, in the case of the monolithic slab system, the foundation and the floor slab are one unit.

The placement of the sand or gravel layer beneath the slab enhances an even curing of the concrete by allowing water to hydrate from the concrete both at the surface and beneath the slab. It is important to moisten the sand or gravel before pouring the slab so the water in the concrete is not pulled into the sand or gravel too quickly. This may lead to lower concrete strengths and excessive cracking. Often a flexible plastic vapor barrier will be placed within the sand to stop moisture migration from the soil up through the slab after the concrete has cured. Steel reinforcement may be used in the slab if significant differential vertical movements are predicted.

A raised floor system consisting of timber or steel joists is another method of supporting the floor. The joists are supported by the foundation system and not the subgrade. Therefore, any predicted movement of the foundation will affect the floor system.

RELATED REFERENCES

ADSC: The International Association of Foundation Drilling and DFI: Deep Foundation Institute, *Drilled Shaft Inspector's Manual,* 1st Edition, ADSC, Dallas, 1989.

American Concrete Institute, *Guide to Residential Cast-in-Place Concrete Construction,* Report No. ACI 332R-84, American Concrete Institute, Detroit, 1989.

Bowles, J. E., *Engineering Properties of Soils and Their Measurement,* 3rd Edition, McGraw-Hill Book Co., New York, 1986.

Brown, R. W., *Residential Foundations - Design, Behavior and Repair,* 2nd Edition, Van Nostrand Reinhold, New York, 1984.

Building Research Advisory Board of the National Research Council, *Criteria for Selection and Design of Residential Slabs-on-Ground,* Report No. 33 to the Federal Housing Administration, National Academy of Sciences, Washington, DC, 1968.

Cole, K. W., *Foundations,* Thomas Telford Ltd., London, 1988.

Das, B. M. *Principles of Foundation Engineering,* PWS Publishers, Boston, 1984.

FLEMING, W. G. K., WELTMAN, A. J., RANDOLPH, M. F., ET AL., *Piling Engineering*, John Wiley & Sons, New York, 1985.

FRENCH, S. E., *Introduction to Soil Mechanics and Shallow Foundations Design*, Prentice-Hall, Englewood Cliffs, NJ, 1989.

GREER, D. M., AND GARDNER, W. S., *Construction of Drilled Pier Foundations*, John Wiley & Sons, New York, 1986.

HOLTZ, R. D., AND KOVACS, W. D., *An Introduction to Geotechnical Engineering*, Prentice-Hall, Englewood Cliffs, NJ, 1981.

INTERNATIONAL CONFERENCE OF BUILDING OFFICIALS, *Uniform Building Code*, ICBO, Whittier, CA, 1988.

NAHB RESEARCH FOUNDATION, INC., *Residential Concrete*, National Association of Home Builders, Washington, DC, 1983.

OLSON, R. E., AND LONG, J. H., "Axial Load Capacity of Tapered Piles," *The Art and Science of Geotechnical Engineering At the Dawn of the Twenty-First Century*, Prentice Hall, Englewood Cliffs, NJ, 1989.

REESE, L. C., AND O'NEILL, M. W., *Criteria for the Design of Axially Loaded Drilled Shafts*, Center for Highway Research, Univ. of Texas, Austin, August 1971.

REESE, L. C., AND O'NEILL, M. W., *Drilled Shafts: Construction Procedures and Design Methods*, for the U.S. Dept of Transportation, FHWA-HI-88-042, ADSC: The International Association of Foundation Drilling, ADSC-TL-4, Dallas, 1988.

SANDERS, G. A., *Light Building Construction*, Reston Publishing Co., Reston, VA, 1985.

SMITH, G. N., *Elements of Soil Mechanics*, 6th Edition, BSP Professional Books, Oxford, 1990.

THOMPSON, H. P., "Supplementary Considerations for Slab-on-Grade Design," *Concrete International*, June 1989, pp. 34–39.

TOMLINSON, M. J., *Pile Design and Construction Practice*, 3rd Edition, Palladian Publications Limited, London, 1987.

ZEEVAERT, L., *Foundation Engineering For Difficult Subsoil Conditions*, 2nd Edition, Van Nostrand Reinhold, New York, 1983.

CHAPTER 4

Examples of Construction at Sites with Expansive Soil

INTRODUCTION

The presence of expansive soil deposits is one of the more challenging site conditions encountered in lightly loaded construction. Cost estimates of damage in the United States to lightly loaded structures due to expansive soil movement range from $660 million in 1973 (Jones and Holtz, 1973) to $1.8 billion in 1979 (Krohn and Slossen, 1980). The dollar amounts become much higher when damage to highways and streets, buried utilities, concrete flat work, and other lightly loaded construction projects are included. Estimates of total damage due to expansive soil movement in the year 2000 is predicted to be $4.5 billion.

> Although earthquakes, floods, tornadoes, and hurricanes can cause much more catastrophic and significant damage, expansive soil movements affect a greater number of lightly loaded structures in the United States.

Although earthquakes, floods, tornadoes, and hurricanes can cause much more catastrophic and significant damage, expansive soil movements affect a greater number of lightly loaded structures in the United States. Jones and Holtz (1973) claim that 14% of the land in the United States is affected by the more catastrophic events listed above during the average American's lifetime and over 20% of the land is affected by expansive soil movements. Chen (1987) points out that structures can now be designed to resist earthquake loads, flood zones have been delineated, and zoning laws now limit what can be constructed within said zones. Finally, tornadoes and hurricanes can be predicted and buildings can be designed to resist them. Expansive soils differ from these other acts of nature in several ways. First, expansive soil movement is not catastrophic; it usually occurs over a relatively long period of time. Second, damage from expansive soils is a continual process. Third, expansive soils cover approximately one-fifth of the United States; therefore, it is impractical to restrict construction in these zones. Finally, expansive soils are locally erratic, making it impractical to delineate levels of damage potential, as is commonly done in earthquake engineering, for example (Chen, 1987).

Although Figure 2.4 shows fairly extensive deposits of potentially expansive soils in the United States, not all of these regions have a high incidence of lightly loaded foundation problems. This is because the climate of the region has a large impact on how much actual swell occurs. The Building Research Advisory Board (1968) published the map of the United States shown in Figure 4.1. The map shows the distribution of a parameter known as the climatic rating (Cw) based on weather data presented by Thom and Vestal (1968). The climatic rating is based on: (1) yearly precipitation, (2) degree of uniformity and distribution of precipitation, (3) frequency of precipitation, (4) duration of each precipitation event, and (5) amount of precipitation for each event. The Cw ranges in value from 15 to 45. In general, a region with inconsistent and/or arid environments will have a low climatic rating. This will typically coincide with a full development of the expansion and shrinkage potential of the clay due to inconsistent weather patterns, causing the soil to go through cycles of shrinking and swelling as the soil dries out and then "takes on" water again. It has been shown that regions with a low climatic rating (below 30) and expansive soil deposits have a higher incidence of distressed structures (Gromko, 1974).

As cities continue to grow and land values increase, pressure to construct upon areas with expansive soil deposits is increasing. Damage amounts in the future may be even higher if design and construction in these areas are not carefully conceived.

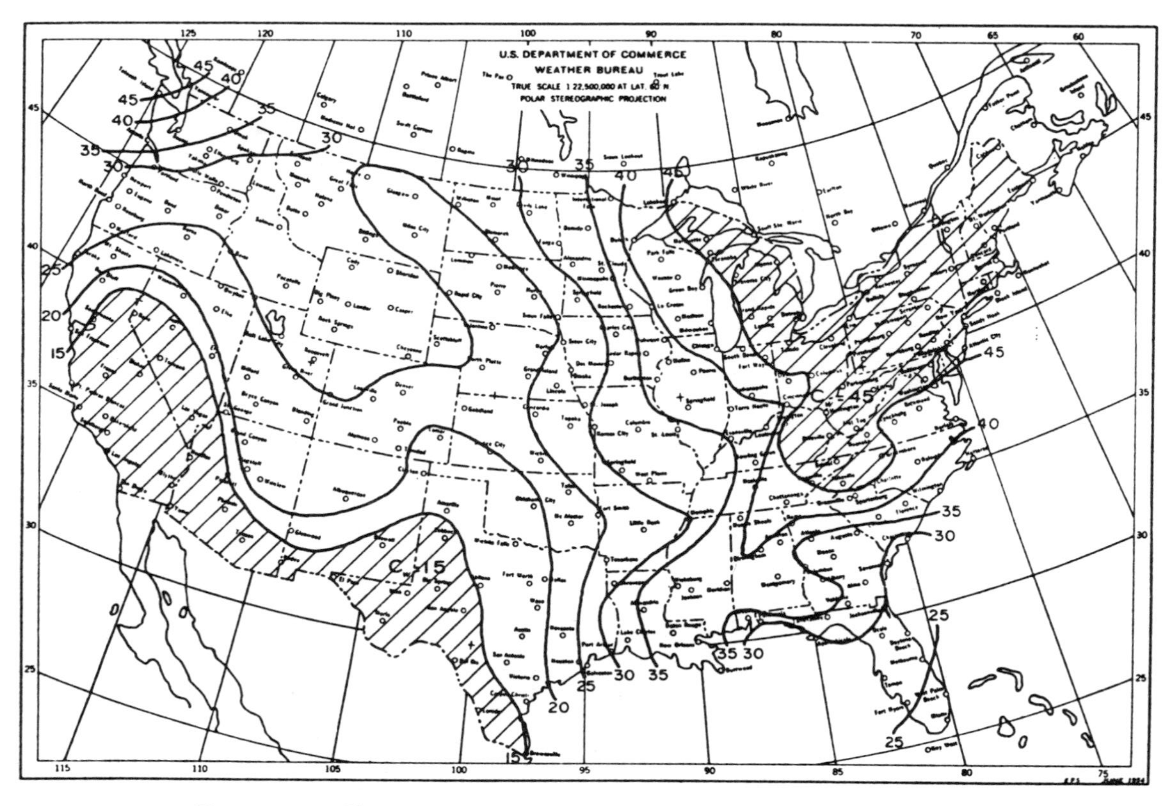

Figure 4.1 Climatic rating (Cw) for the United States. (BRAB, 1968)

As cities continue to grow and land values increase, pressure to construct upon areas with expansive soil deposits is increasing. Damage amounts in the future may be even higher if design and construction in these areas are not carefully conceived.

The majority of the more challenging site conditions that may be encountered present problems concerning the design of foundation systems for control of settlement. However, the presence of expansive soils presents the exact opposite challenge: that is, how to hold a building down. Lightly loaded structures simply do not provide enough downward load to offset the high uplift swell pressures that can occur. Swell pressures as high as 20,000 psf, which far exceeds the typical design load of a lightly loaded structure, have been measured (Holtz and Kovacs, 1981). These large swell pressures generally are not uniform, which leads to large differential movements. It is the differential soil movements that cause the most damage to structures.

Four general construction techniques are used to reduce the potential for damage to a structure from expansive soil movement. First, hold the structure down by utilizing a deep foundation system founded in soils that are either nonexpansive or are believed to be unsusceptible to seasonal moisture variations. This is most commonly done using a drilled shaft foundation system. Second, internally strengthen the foundation to make it strong enough to undergo the predicted differential movements without cracking and without imparting damage to the structure. Systems that are typically strengthened include isolated and continuous footings, monolithic slab foundation systems, and post-tensioned slabs. Post-tensioned slabs are discussed on page 60 of this chapter. Third, remove the expansive soil to a depth of predicted seasonal moisture variation and replace it with nonexpansive select fill. Fourth, treat the subgrade soil to decrease its expansion potential.

This chapter examines the state of the practice of construction of lightly loaded foundations on expansive soils in different regions within the United States, including Denver, Dallas, Southern California, and the San Francisco Bay area. The purpose is to present more than one solution for a similar challenging site condition—expansive soil. Advantages and disadvantages and common design and construction oversights related to different foundation systems for lightly loaded construction at sites with expansive soil deposits are summarized at the end of the chapter.

DENVER, COLORADO

Area Description, and Geologic and Soil Conditions

The Denver area is characterized by gently rolling plains toward the east and a more rugged terrain toward the west in what constitutes the eastern foothills of the Rocky Mountains. The climate is relatively arid. The climatic rating (Cw) for the region is approximately 22. This implies that the overall weather pattern throughout the year is relatively inconsistent, leading to large variations in the near-surface soil moisture contents.

The predominant near-surface geologic formations are the Pierre, Denver-Dawson, and Laramie shales in the plains and the Mancos and Lewis shales in the foothills. These formations are commonly underlain by bedrock known as the Denver Formation (Taylor, 1990). All of these formations contain significant amounts of montmorillonite clays and therefore have a high potential for expansion. Swell pressures as high as 15,000 psf and vertical swells of 2 to 4 inches are common. Vertical swells greater than 4 inches are not unheard of.

Predominant Foundation System

Drilled shaft systems in Denver. Engineers and builders in the greater Denver metropolitan area have consistently used drilled shaft foundation systems since the 1950s to minimize damage from expansive soils. The overall design concept is to drill the shafts into the zone of soil that is not expected to undergo significant moisture variations. Penetration into this zone should be deep enough to develop enough friction between the shaft and the soil to offset the tendency for the expanding soil to lift the shaft. A diagram illustrating these offsetting forces is shown in Figure 4.2.

The capacity of the shafts is derived from both end bearing and side friction for vertical compressive support. Bearing capacities of the soils in the area are usually relatively high and, therefore, the controlling parameter of the design depth is dependent on developing enough

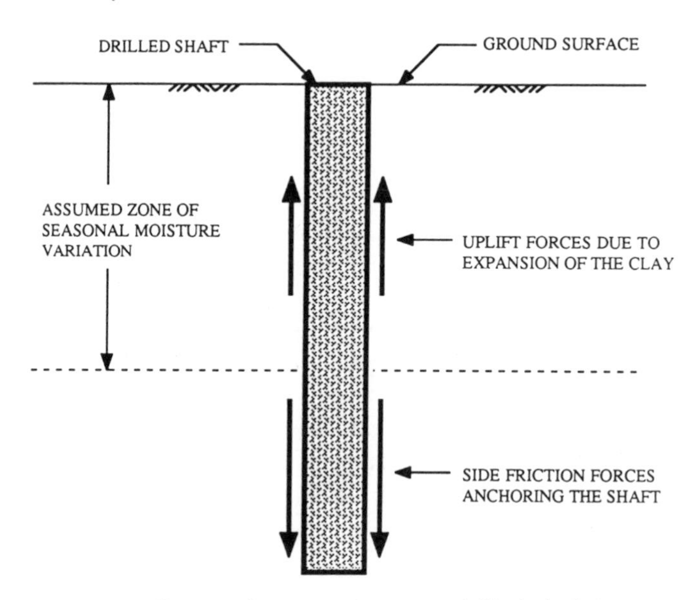

Figure 4.2 Offsetting forces acting on a drilled shaft in expansive soil.

side friction to offset the potential uplift forces in the zone of moisture variation.

The shafts are typically straight, 15 to 20 feet in length with 10- to 12-inch diameters. Steel reinforcement usually consists of a minimum of two full-length, Grade 40, No. 5 bars. The shafts are typically drilled with truck-mounted rigs with a total reach of 21 feet, as shown in Figure 4.3. If it is necessary to construct a deeper shaft, a larger drill rig must be mobilized. If possible, design depths are kept within 20 feet so as to use the less expensive smaller rig.

It is becoming common to construct shear rings (Figure 4.4) in the lower portion of the shafts to increase the "hold-down" effect. The shear rings are created by attaching a spring-loaded grooving tool to the drilling auger once the shaft depth has been reached.

The diameters of the shafts are kept small to reduce the surface area with which the soils in the zone of moisture variation are in contact, thus reducing the potential uplift forces. The steel reinforcement is necessary to hold the shaft together when it is subjected to tensile forces from the swelling of the expansive soils. Spacing between the shafts is typically maximized. This minimizes the number of shafts that are necessary, and increases the dead load on each shaft. The increased downward

Figure 4.3 Truck-mounted drill rig drilling a small diameter shaft in Denver, Colorado. The drilled shaft is being constructed at the basement grade.

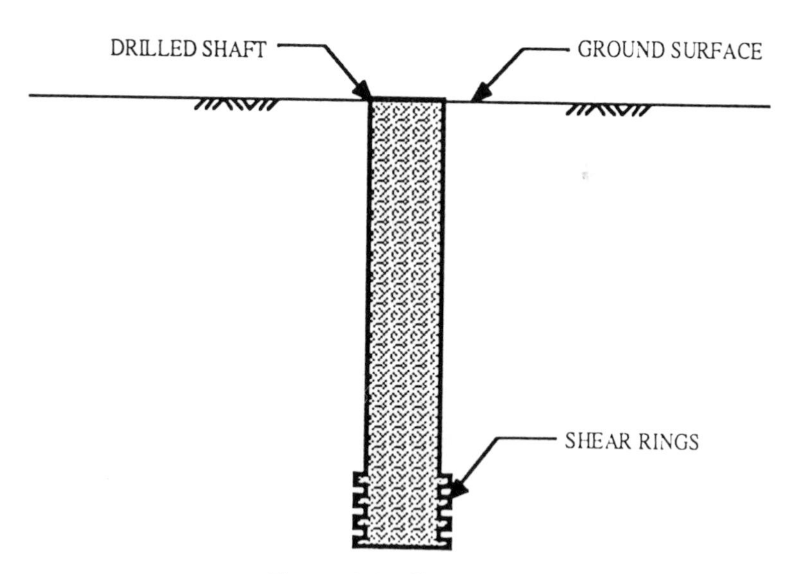

Figure 4.4 Shear rings.

load on each shaft helps to offset some of the upward forces due to the zone of expansive soils. A rule of thumb is that it is economically viable to increase the shaft spacing until it becomes necessary to use stirrups in the grade beams to span the spacing. Stirrups are horseshoe-shaped reinforcing steel members that strengthen a concrete beam against shear failure.

Grade Beam Construction. Basement construction is common in the Denver area. Therefore, 8-inch-thick basement walls usually serve as the grade beams. The basement floor slabs are typically constructed on the subgrade and are "free floating" to enable movement of the slab without any stress being transferred to the structural members of the building. Any moisture that enters the basement between the walls and slab is directed toward a sump in the basement and pumped off site. *As Chen (1988) states, "Few realize that there is no such thing as a truly floating slab.* All slabs-on-ground adhere to the grade beams . . . [transferring] pressure to the grade beam affecting the structural stability." The cost-savings of the slab-on-grade versus a raised structural floor is substantial enough that many builders will accept the associated risk of damage caused by slab heave.

For buildings without basements, the grade beams are typically 36 to 42 inches deep and 8 inches wide. Most of these buildings also have floating floor slabs. However, there is a local trend, especially for upscale construction, to use a timber (residential) or steel (commercial) floor system with a crawl space beneath it. Although this method is more expensive, it is advantageous because if properly designed and constructed, the floor will not be subjected to any expansive soil movements (assuming the foundation remains stable).

One of the more important aspects of the design is a void that is created beneath the grade beam (or basement wall). Typically in Denver, a 6-inch void is left beneath the grade beam to allow the soil to expand without coming in contact with the beam. The void is created by placing a corrugated cardboard void box on the subgrade in the bottom of the forms. The top and bottom of the void box is manufactured with a polyurethane coating to keep the water in the concrete and soil from disintegrating the cardboard immediately. The sides of the box are left untreated. Once the forms are removed and the wall or grade beam is backfilled, moisture from the backfill weakens the sides of the cardboard void box. This effectively creates the void be-

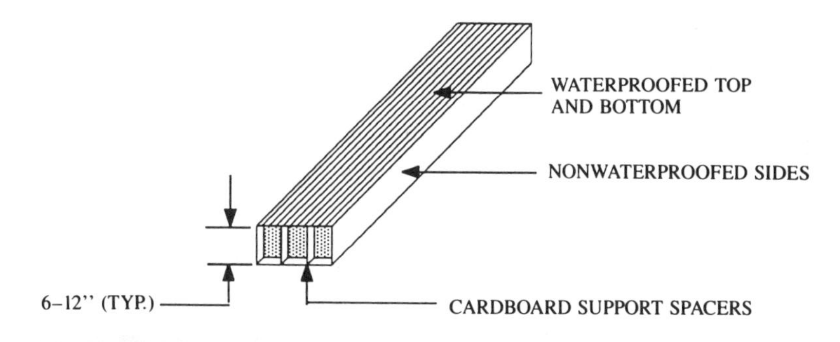

Figure 4.5 Typical void box construction. Photograph shows void box in place beneath a crawl space grade beam.

tween the soil and the grade beam. Figure 4.5 illustrates a typical void box and contains a photograph of a void box beneath a crawl space wall.

Other Foundation Systems in Denver

Several other foundation systems or techniques have been used in the Denver area with varying degrees of success. Removal of the upper 3 to 4 feet of expansive material and replacement with a nonexpansive material is one method. The design concept is that if the moisture variation depth is only 3 to 4 feet, the soil that has potential for expansion will not be subjected to a moisture variation. Once this has been done, the foundation of choice is usually a continuous wall footing. This method is not generally accepted in the Denver area, however. Many believe this procedure can create a "bathtub" beneath the foundation. The nonexpansive material is generally more permeable than the expansive clay soils. Therefore, the water is able to seep into the replacement soil and is then perched on top of the clay soil that was not removed. This perched water then becomes a source of moisture, causing the remaining clay to expand.

Another system used in the Denver area is the post-tensioned slab foundation. Post-tensioned systems consist of slabs reinforced with high-stress steel strands that are tensioned after the concrete has begun to cure. The advantage of this system is that it creates a more rigid slab that can withstand greater differential movements without cracking. It also reduces the potential for concrete shrinkage cracks to open up and thus eliminates the need for expansion joints. Details of the post-tensioned system are presented in the section on Dallas, Texas.

During the mid-1980s the post-tensioned slab foundation system was fairly prevalent at sites with a high expansion potential. Its use, however, has not remained popular in the region. This is possibly because of: (1) an unfamiliarity with the design and construction procedures resulting in a relatively high failure rate, (2) the potential for the foundation to remain intact but move out of level, requiring future repair, (3) difficulties related to construction procedures at the basement level, and (4) difficulties related to any necessary future repair work involving cutting through the slab without damaging the post-tensioned strands.

Site Drainage

The presence of expansive soil deposits in Denver makes site drainage a very important consideration. Any surface water allowed to seep below the foundation could cause the subgrade soil to expand. The drainage design should be such that a relatively constant moisture content is maintained in the subgrade soil.

Several techniques are used to control the drainage. One technique is to compact the backfill material around the perimeter of the building at a moisture content above the optimum moisture content. As discussed in Chapter 2, compacting at moisture levels above the optimum moisture content helps to reduce the expansion potential and the permeability of the backfill. Another technique is to make sure that the backfill slopes away from the building on all sides. This enables the transport of any surface water away from the foundation before it has a chance to seep into the subgrade.

Subdrains (perforated pipe or free-draining gravel) may be installed at the same elevation as the bottom of the void box around the perimeter of the building. Thse drains transport any water that seeps into the subgrade to a sump in the basement, preventing any swelling of the soil beneath the foundation and slab. Subdrains are somewhat controversial, however. Some designers feel that if they are not properly installed, the presence of the free-draining gravel can act as a conduit to transport the water to the clay beneath the foundation and slab, thus increasing the potential for soil movement. It is recommended that the subdrains be constructed with a layer of impervious plastic on the building side of the drain to help prevent this "conduit" effect.

Careful attention is also paid to roof drainage. Many of the newer buildings in the Denver area have extensive roof drain systems, ensuring that all of the precipitation falling on the roof is controlled. Outlets of the drain pipes are commonly extended at least 5 feet away from the building, as shown in Figure 4.6, to keep the roof runoff away from the foundation.

Extremely Difficult Conditions

Some site conditions in the Denver area are considered to be extremely difficult and risky to build upon. For example, near the Rocky Mountain Range the clay strata have been tilted at angles of 60°–70°. The tilting was caused by the original uplift of the mountain range. These clay strata are also highly fissured or cracked. It is believed that the tilting of the clay strata and the presence of the fissures enables surface water to penetrate relatively deeply. Traveling along the fissures, it is thought that the surface water may reach depths of 30 to 40 feet below the ground surface.

There have been numerous distress problems in this region even with drilled shaft foundations reaching depths up to 20 feet below grade. It is now believed, locally, that it will be necessary to drill shafts 35 to 40 feet deep in these regions to minimize the chances of structural distress. It is not known at this time whether this will be effective; however, drilled shafts remain the most effective and economical system in

Figure 4.6 Extended roof drain in Denver, Colorado.

the region at sites with expansive soil. Others in the area feel that it is simply too risky to construct lightly loaded structures on site conditions of this nature. The desire to build in these locations is strong because of spectacular views and, as stated in Chapter 1, most of the "good" sites already have structures on them.

DALLAS, TEXAS

Area Description, and Geologic and Soil Conditions

The greater Dallas area is characterized by fairly flat or gently sloping plains. Similar to Denver, the climate is relatively arid. The climatic rating for the region is approximately Cw = 20. This implies that the overall weather pattern throughout the year is very seasonal. Large amounts of rain may fall in the winter for short periods of time, yet the summers can be quite dry. This pattern leads to large variations in the amount of moisture in the near-surface soils throughout the year.

Two of the more predominant, near-surface geologic formations are the Eagle Ford shale and the Woodbine shale. Both formations contain significant amounts of montmorillonite clays and have a high potential for expansion. Swell pressures as high as 20,000 psf and vertical swells of 5 to 6 inches are common; vertical swells of 8 to 12 inches have been measured.

Predominant Foundation Systems

Two different general foundation systems are commonly used in the Dallas area to minimize the potential damage caused by expansive soil movement. First, variations of the monolithic, thickened edge, reinforced slab-on-grade system are used at sites with expansive soil deposits. Second, similar to the Denver area, drilled shaft foundation systems are utilized for lightly loaded structures constructed on expansive soil. The choice between these two systems depends on a variety of factors including: (1) the expansiveness of the soil, (2) the builder's budget, (3) the structural tolerance of the proposed building, and (4) the personal preference of the builder, contractor, or engineer. In general, the drilled shaft foundation is more commonly used for commercial structures and the slab-on-grade foundation is more commonly used for tract homes. Neither system is used exclusively for one of the above classes of construction; however, drilled shafts typically enter the residential market in Dallas only for upscale custom homes. The drilled shaft system is more common on commercial structures because the budget is usually larger and can accommodate the more expensive system. In general, if properly constructed and designed, the drilled shaft foundation will not undergo as much movement as the slab-on-grade system. For commercial structures the drilled shaft system is worth the extra expense because the builder's intent is generally long-term, self-ownership of the structure. Conversely, residential structures are typically sold immediately to individual owners. The builders therefore attempt to keep budgets at a minimum so their product is affordable to the buyer.

Drilled shaft systems in Dallas. The design concept of the drilled shaft system in Dallas is similar to that in Denver. The shafts are drilled deep enough to provide enough downward side friction along the shaft to offset the upward forces generated from expansive soil movement caused by changes in seasonal moisture near the ground surface. The shafts can be either straight or belled depending on the strength of the bearing stratum. Bells are commonly used to reduce the contact pressure by spreading out the column loads.

The depths of the shafts can vary from 5 to 25 feet depending on the predicted depth of the zone of seasonal moisture variation and the depth to bedrock. The deeper designs generally have called for embedment into the bedrock, thus requiring additional drilling depth, yet adding to the factor of safety against movement.

Shaft diameters range from 12 to 18 inches for residential construction and 12 to 24 inches for commercial structures. The diameters are typically larger for the commercial structures because of higher column loads. For belled shafts it is necessary to have a minimum 12-inch diameter to fit the belling tool in the shaft. Bell diameters vary depending on the bearing stratum, but it is recommended not to exceed three shaft diameters (Reese and O'Neill, 1988). A shaft 12 inches in diameter[1] is the smallest that local contractors feel is "workable" in terms of steel and concrete placement, and this same diameter is generally required for bearing purposes.

Reinforcing steel is commonly placed along the full length of the shaft to provide the required tensile strength to offset uplift forces in the zone of seasonal moisture variation. Two No. 6 reinforcing bars are the most common design, ranging up to a maximum of two No. 9 bars. Grade beams are typically reinforced with two No. 5 reinforcing bars near the top and near the bottom of the beam.

Spacing between the shafts will vary from 10 to 15 feet. It is desirable to maximize the spacing to increase the dead load on each shaft without exceeding the capacity of the bearing layer, or requiring massive grade beams to support the span.

Cardboard void boxes are commonly used and range in height from 6 to 14 inches. Basement construction is not common in the Dallas area. Therefore the void box is typically placed near the final grade elevation. It is recommended that the void box be placed slightly below finish grade. If it is placed at finish grade, when it deteriorates there is an open conduit for water to travel under the house and cause the soil to swell. By placing it below finish grade the potential for surface water reaching the subgrade beneath the house is greatly reduced because it will first need to seep through the compacted backfill around the perimeter grade beams and then flow past any subdrains that may have been installed.

The floor systems for drilled shaft foundations in the Dallas area consist of either a raised structural floor made of timber or steel, or a

[1]Although 10-inch diameter shafts are frequently constructed in Denver, they are rarely constructed in Dallas.

free-floating slab-on-grade. It is more desirable to use a raised floor system, albeit more expensive. Typically, if a floor slab is resting immediately on the subgrade, a method of subgrade treatment will be used to reduce the expansion potential of the near surface soil. These methods include pressure injection of water, chemical injection, lime slurry pressure injection, or lime addition. These methods are discussed in further detail in the following section on slab-on-grade systems. However, if these treatment systems are not effective, the combination of a drilled shaft system and a slab-on-grade floor may result in significant differential movements if the slab heaves and is not completely detached from the grade beams.

Monolithic reinforced slab-on-grade foundation systems in Dallas. The two different types of monolithic slab-on-grade foundation systems commonly used in the Dallas area are: (1) conventionally reinforced and (2) post-tensioned. Both types of slabs are commonly designed with deepened stiffening beams running both directions in the slab. When deepened beams are used to add additional stiffness with the conventionally reinforced slab, it is commonly referred to as a "waffle" slab.

The decision between using a conventionally reinforced or a post-tensioned reinforced slab depends on: (1) the expansiveness of the soil, (2) the builder's budget, (3) the structural tolerance of the proposed building, and (4) personal preference and familiarity. Each system can be built to support the building and maintain its integrity for a variety of expansive conditions. However, as the expansion potential becomes more severe it may become economically beneficial to choose one system over the other.

The design concept of these systems is to reinforce them in such a way that they are rigid enough to move as a single unit in the event of differential soil movement. If properly reinforced and constructed, differential movement should not impart damaging stresses to the structure itself. Although these systems are able to resist some movement, the possibility exists that if there are large differential movements, excessive tilting may occur and it may be necessary to relevel the foundation and the building. Many builders are willing to risk future remedial work and use one of these less costly foundation systems.

Conventionally Reinforced Slabs. The conventionally reinforced slab has been a common type of lightly loaded construction since the end of World War II (Post-Tensioning Institute, 1989a). In 1957 the Building Research Advisory Board (BRAB) began a research project to develop criteria for the selection of a slab type for different soil conditions. The research was concluded in 1968 at which time *Criteria for Selection and Design for Residential Slabs-on-Ground* (BRAB, 1968) was published. The report identifies our different types of slabs:

Type I	Unreinforced
Type II	Lightly reinforced against shrinkage and temperature cracking
Type III	Reinforced and stiffened
Type IV	Structural (not directly supported on the ground)

The type that is of interest here is Type III. A typical design in Dallas may consist of a 4- to 5-inch-slab, reinforced with No. 3 reinforcing bars 16 to 18 inches on center. Designs will vary depending on the magnitude of the potential soil movement. The "stiffened" portion of the slab refers to the waffle-type construction, where deepened beams within the slab increase its stiffness. Refer to Figure 3.3 for a typical cross section of a waffle slab with reinforcement.

For design and construction recommendations for a slab-on-ground, readers should refer to the above-referenced BRAB publication; also see the American Concrete Institute report ACI 332R-84 entitled *Guide to Residential Cast-in-Place Concrete Construction,* published in 1989; the National Association of Home Builders' (NAHB) publication entitled *Residential Concrete,* published in 1983; and UBC Standard 29-4, Part I, 1988 edition (ICBO, 1988).

Post-Tensioned Reinforced Slabs. Post-tensioned reinforced slabs have become very popular in the Dallas area since their initial introduction in the mid-1960s. The main advantages of the post-tensioned slab over a conventionally reinforced slab is the elimination of joints and cracking by installing post-tensioning strands in the slab as the reinforcement. Figure 4.7 is a diagram of a typical post-tensioning strand layout, and Figure 4.8 shows a photograph of a similar layout.

Unbonded strands are typically used for lightly loaded slab-on-grade applications. An unbonded strand consists of a bundle of seven individual steel strands that have been twisted together, greased, and coated with plastic. Figure 4.9 shows a closeup of an installed strand.

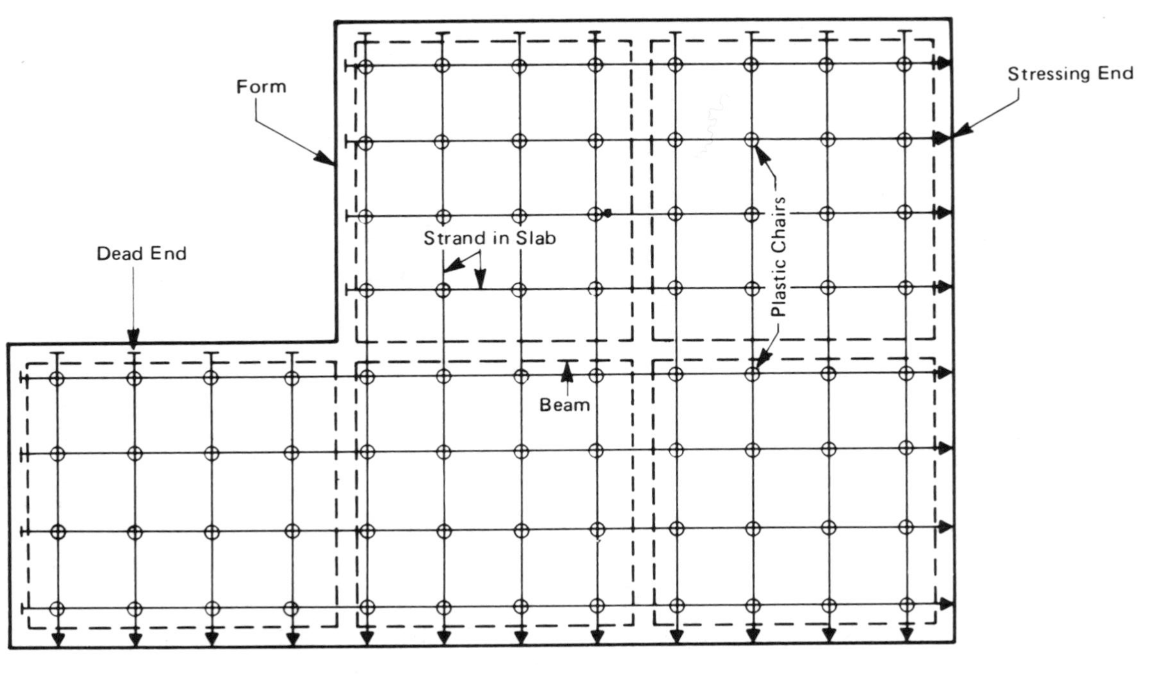

Figure 4.7 Typical post-tensioned slab layout. (Reprinted by permission of the Post-Tensioning Institute, PTI, 1989.)

Figure 4.8 Photograph of a typical post-tensioned slab layout.

The plastic coating around the strand can be seen in the lower left corner of the photograph. The seven strands typically have a combined diameter of 0.5 or 0.6 inches and the ultimate tensile strength of the steel if 270,000 psi. Each end of the strand is attached to a steel plate measuring approximately 2.5 × 5 inches, using split conical-shaped steel wedges or collets that grip the cables as they are inserted into the plate. The plate can be seen in Figure 4.9. Each strand has a "dead" end and "live" or stressing end.

Unlike conventional steel reinforcement, the strands are relatively flexible and are delivered to the job site rolled up. The construction procedure entails rolling out and then chairing up the strands similar to conventional reinforcement. Figure 4.10 shows a tendon being rolled out into position. The dead ends are placed a minimum of one inch inside the edge of the slab. Figure 4.11 shows the dead end of a tendon temporarily attached to the concrete form. A pocket former is attached to each of the live ends, which are then attached to the forms. The pocket former abuts the concrete forms at the live end as shown in Figure 4.9.

The forms are removed one day after the concrete pour so the pocket formers can be removed before the concrete becomes too hard.

Figure 4.9 Typical "live" end of post-tensioning strand. Note the plastic sheathing around the bundle of individual strands, the metal plate and the pocket former abutting the concrete forms.

The purpose of removing the pocket former is to create room to insert the tensioning equipment into the slab to grip the tendons. Once the concrete attains a minimum compressive stress of 2000 psi (usually 3 to 10 days for most concrete mixtures), the tensioning of the strands is performed.

The grease and plastic around the strands isolate them from the concrete. This enables the strands to be tensioned (stretched) without bonding to the concrete and pulling the slab apart. The strands are elongated when they are tensioned and are held in this position by the steel wedges. The steel wedges transfer the tension in the strands to the steel plates at each end of the strand. The elongated strands force the steel plates to compress the concrete, creating a more rigid slab that will have less tendency to crack. In essence, the concrete is being "held" together by the stressed tendons.

In Dallas, a post-tensioned slab is typically 4 inches thick with the strands placed in both directions. The spacing between strands ranges from 4 to 5 feet. Depending on the expansiveness of the soil, the

Figure 4.10 Post-tensioning strand being rolled out into its position within the slab.

post-tensioned system can become cost-competitive with a conventionally reinforced system because the conventional system requires a thicker slab and tighter spacing of the reinforcement (16 to 18 inches apart) to be effective on highly expansive soils.

Readers are encouraged to refer to the Post-Tensioning Institute's *Design and Construction of Post-Tensioned Slabs-on-Ground* (PTI, 1989a), as specified by UBC Standard 29-4, Part II, 1988 Edition (International Conference of Building Officials, 1988), for design purposes.

Waffle Slab. In Dallas, when soils are expected to undergo large amounts of movement due to expansion even after subgrade treatment, waffle-type slabs are sometimes used. As mentioned above, waffle slabs consist of conventionally reinforced slabs with deepened reinforced beams running both directions in the slab. The deepened beams stiffen the slab such that the slab can withstand larger uplift pressures than a uniformly thick slab. A trencher is usually used to excavate the subgrade in order to pour the stiffening beams.

Subgrade Treatment. It is common to use some method of subgrade treatment in an attempt to reduce the potential for expansion of the

Figure 4.11 Typical "dead" end of post-tensioning strand attached to concrete form.

soil beneath the slab. These methods include water pressure injection, chemical injection, lime slurry pressure injection, or lime addition. These methods are described below.

Water pressure injection increases the moisture content and preswells the soil. The floor slab should be poured soon after the injection to ensure that the slab will cap the subgrade and thus retain its high moisture content. If the slab is not poured soon after the injection, the subgrade will dry out and the potential for expansion will remain high. According to Jones and Jones (1988) this method can be ineffective because the water may not be capable of rapidly penetrating the clay. Rather, it may just travel along fissures or more permeable lenses in the clay deposit.

Chemical pressure injection consists of injecting chemicals into the soil that will chemically alter the clay minerals, rendering it nonexpansive. At least 85% of the potentially active depth of the expansive material should be treated for the procedure to be effective (Jones and Jones, 1988). Successful applications may be difficult to apply effectively in nonhomogeneous clay deposits because varying amounts of chemicals will be required.

Lime slurry pressure injection has been demonstrated to be effective in reducing the amount of moisture migration through the soil (Boynton and Blacklock, 1985; Wright, 1973, 1974). Lime is added to water and injected into the soil. It is thought that the slurry coats the clay where it has penetrated the soil and inhibits additional moisture from swelling the coated particles. However, the effectiveness of the lime slurry uniformly coating the clay is debatable. It is the authors' opinion that the injected lime slurry would fill fissures or penetrate relatively weak zones in the clay, thus isolating only certain zones within the expansive subgrade. In addition, the long-term effectiveness of this technique has not been universally proven (Jones and Jones, 1988).

Lime slurry pressure injection applications typically range from 7 to 10 feet deep on an average horizontal spacing of 5 feet. The slurry is injected into the soil every 8 to 12 inches of depth (Transportation Research Board, 1987). It is recommended that the upper 12 inches of soil be stabilized with conventional methods of mixing lime into the soil and recompacting it.

The mixing of lime with the soil reduces the expansion potential by chemically interacting with the clay minerals. The addition of lime also increases the workability of clay soil and makes grading the soil easier. This method typically is cost-effective only where large-scale grading operations are already planned before the option of lime addition is chosen. This is because a large amount of time and effort is required to thoroughly mix the lime with the clayey soil to render the procedure effective. In addition, lime-treated soil tends to be susceptible to a loss of effectiveness during and after freeze/thaw cycles and thus may not be as effective in colder climates (Transportation Research Board, 1987).

In addition to these methods it is also common to use a layer of sand or gravel beneath the slab, which has been placed on the subgrade. The sand or gravel layer reduces the potential for shrinkage cracking during the curing of the concrete by providing drainage beneath the slab for the water hydrating from the concrete. This creates a more uniform curing rate within the slab. Often a flexible plastic vapor barrier will be placed within the sand or gravel to stop moisture migration up through the slab after it has cured. The moisture barrier helps to keep moisture-sensitive floor coverings dry. It also keeps the subgrade soil from drying out and shrinking during a drought year by stopping transpiration through the slab. It is recommended that the sand layer be moistened immediately before the slab is poured. If the sand is dry it will rapidly draw moisture from the concrete, which will cause the concrete to cure too quickly, resulting in excessive shrinkage cracks.

Site Drainage

The presence of soils with a high potential for expansion necessitates that attention be given to surface water drainage. Attention is paid to roof drainage as well as the construction of slopes leading away from the foundation. Perhaps more importantly, a typical specification in Dallas may include a 10-mm. plastic liner attached to the outside edge of the grade beam or slab, extending a minimum of 6 feet from the grade beam and sloping away from the structure. The plastic liner is then covered with landscaping soil. The plastic liner creates an impermeable surface that channels water away from the foundation, reducing the potential for expansion of the subgrade soils beneath the foundation.

Another method used in Dallas that serves the same function as the plastic liner is to place compacted clay around the perimeter of the foundation. The clay layer should be compacted above the optimum moisture content and slope away from the foundation. As discussed in Chapter 2, compaction at moisture levels above the optimum moisture content causes the clay to be more impermeable and have a lower potential for expansion. However, if the compacted clay dries out, it will lose some of its effectiveness and possibly undergo some shrinkage. This may not be as permanent a solution as the plastic liner.

Extremely Difficult Conditions

Similar to Denver, there are some challenging locales in Dallas with highly swelling clays. Swells over 12 inches in vertical rise have been observed in the Las Colinas area. One of the recurrent observations regarding this area and other highly swelling deposits is that many engineers tend to rely too heavily on empirical data rather than their own better judgment and/or laboratory tests. An example of this is the correlation between the plasticity index and the percent volumetric change (Texas Highway Department, 1972). Once the percent volumetric change is known, the potential vertical rise (PVR) can be estimated. This relationship has been successfully used for many projects. Recently, outlying areas of Dallas have been developed where clays with critically high expansion potentials exist. In hindsight it has been discovered that the

PVR–PI relationship has not been valid for these soils. Projects that have been constructed using this relationship have exhibited significant distress. It is important to remember that empirical relationships provide excellent guidelines, but site-specific information is invaluable in determining the applicability of the empirical relationship in question.

SOUTHERN CALIFORNIA

Area Description, and Geologic and Soil Conditions

Southern California is characterized by numerous mesas, canyons and valleys, and mountain ranges toward the east. The majority of the soils in the region are silty and sandy. The foothills of the mountains typically consist of soil comprised of decomposing granitic and/or volcanic rocks underlain by the parent formation. The volcanic rocks may commonly decompose to soils with a high expansion potential. There are also geologic formations of marine origin that possess significant expansion characteristics. These expansive clay deposits are most prevalent in northern San Diego County and Orange County.

The climatic rating (Cw) for the region is 15, which is the lowest rating possible. This indicates that the weather pattern is very seasonal with long periods of little or no precipitation. This will lead to periodic changes in the moisture content of the near-surface soils, which, as discussed, may lead to significant volume changes in the expansive clay deposits.

Predominant Foundation Systems

Continuous wall and isolated spread footings in Southern California. In general, this system, built in conjunction with a concrete floor slab-on-grade, is undoubtedly the most common system in the Southern California region. Because of the popularity of this system, attempts are made to use it even when confronted with a site with expansive soils. Laboratory tests are utilized to determine the expansion index (UBC Standard 29-2), (ICBO, 1988) of the upper 3 feet of soil, which corresponds to the typical predicted depth of moisture variation. The UBC expansion index is a measure of the percent increase in the change in height of a sample upon saturation under loading of 1 pound per square inch (psi). The expansion index is scaled by a factor of 10. That is, an expansion index of 90 indicates a percentage increase of 9%. When expansion index values are greater than 90, the soil is considered to be highly expansive.

Once the expansion potential is known, the general approach is to increase the amount of reinforcement in the continuous footings and slabs as the potential for expansion increases. This, of course, makes the continuous footings and slab more rigid and reduces the potential for cracking due to differential movement and/or a loss of structural integrity. The amount of extra reinforcement required is based on empirical results. Increasing the amount of reinforcing steel in isolated spread footings would reduce the potential for cracking of the footing itself, but it would not decrease the potential for differential movement across the length of the building.

The typical designs range from 12-inch deep footings with a 12-inch-wide base and one No. 4 bar in the top and bottom of the footing for a structure on a subgrade with a low expansion potential, to 24-inch-deep footings with a 12-inch-wide base and two No. 4 bars in the top and bottom of the footing for a structure on a subgrade with a high expansion potential. The slab sizes will range from 3.5 to 5 inches thick. Recently, 3.5-inch-thick slabs have become less common, especially for slabs on expansive soil. Reinforcement of the slabs typically range from 6 × 6 10/10 wire mesh to No. 3 steel reinforcing bars at 18 inches on center, both directions, for sites with highly expansive subgrades. Wire mesh consists of cold drawn wires laid out in a grid, which have been welded at each of the intersections. It is usually specified in terms of the spacing between the wires and the cross-sectional area of the wires, in hundredths of a square inch. The specification listed above (6 × 6 10/10) refers to a 6-inch spacing in both directions, using wires with a cross-sectional area of 0.1 square inches, in both directions. Welded wire mesh helps to retard the widening of cracks by interconnecting the concrete on either side of a crack; however, it does not add any significant flexural strength to the slab.

In general, the floor slab subgrade is either presoaked by flooding to increase the moisture content above optimum, or particular attention is paid to ensuring that the subgrade moisture content is maintained at a constant level beneath the slab. Traditionally, the slabs have not been structurally designed. It has been assumed that the soil will support the slab and the steel will offset any problems related to shrinkage cracks in the concrete. This does not guarantee that there will not be any cracking of the slab because of differential movement of the soil. High land values in Southern California have caused individual homeowners to become extremely sensitive to slab movements and cracking. Litigation is very common even for shrinkage cracks that do not necessarily affect or threaten the structural integrity of the house. It is difficult to explain to a homeowner who has just invested his or her life's savings into a home that concrete will naturally crack when it shrinks while curing, but without any loss of structural integrity.

The number of construction defect lawsuits has increased dramatically in this region during the last 10 years. This has forced engineers and designers to use a methodology that incorporates a rational structural design of the slab based on predicted soil movements and corresponding stresses in an attempt to reduce potential cracking. Unfortunately, many soil reports do not typically provide some of the parameters necessary to do this. An example of this is the parameter "edge moisture variation distance." This parameter predicts how far from the edge of the slab the moisture content of the soil will vary significantly owing to changes in the weather. That distance, as well as the magnitude of potential vertical movement of the soil, is important in determining the magnitude of the stresses induced in the slab because of differential movements across it.

In Southern California, the desire to control cracking of the slabs and differential movement of the footings, and to use a system that is not empirical, has led to an increase in the use of post-tensioned slab foundation systems for construction on expansive subgrades.

Post-tensioned slab foundations in Southern California. Despite local problems of unfamiliarity with design procedures and the fact that post-tensioned slabs are more construction-sensitive, post-tensioned slabs are becoming more and more common. Developers are willing to spend the additional money to reduce the potential for cracking in the slab, possibly reducing the number of lawsuits. As the potential for expansion increases, the post-tensioned slab becomes cost-competitive with continuous footings because of the increased steel and depth required for an appropriately designed footing system.

The thicknesses of these slabs typically range from 4 to 5 inches and the strands are spaced 4 to 5 feet apart in both directions. In general, the edges of the slab are thickened to 12 inches. There have been some cases of slabs curling up at the edges from heaving soil, as shown in Figure 4.12. Therefore, many designs now call for an 18-inch deep thickened

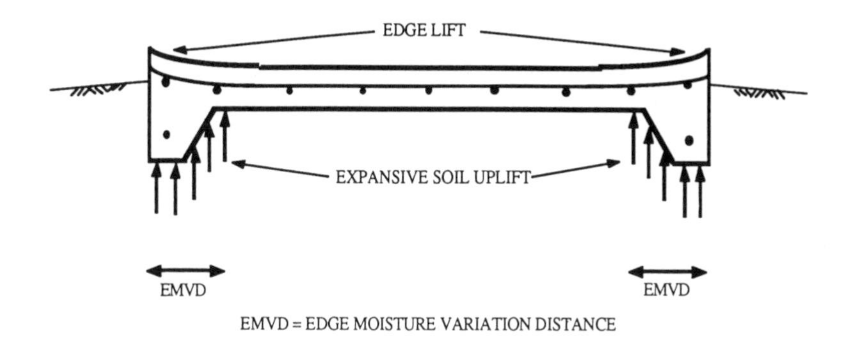

Figure 4.12 Edge lift of a post-tensioned slab.

edge to reduce the edge moisture variation distance in an attempt to curtail the uplift forces at the edges of the slab.

Some of the local oversights that have occurred with the post-tensioned design are related to an unfamiliarity with the design. Geotechnical engineers do not routinely provide the necessary parameters outlined in the Post-Tensioning Institute's design method such as the edge moisture variation distance, the predicted differential soil movement based on the potential swell of the clay, the slab-subgrade friction coefficient, or the mineralogy of the potentially swelling clay. The slab-subgrade friction is the force that resists the movement (compression) of the slab during stressing of the tendons. The tendons must be spaced closely enough to overcome the slab-subgrade friction. Therefore, the structural engineer may specify a slab design that local post-tensioned slab companies have designed. Typically these slabs are designed strictly to be cost-competitive with a conventionally reinforced slab. Thus, cost controls the design of the post-tensioning system, which, depending on the expansion potential of the soil, may or may not produce an adequate system.

At present some controversy surrounds the necessary design requirements in the Southern California region. The latest version of the *Uniform Building Code* (International Conference of Building Officials, 1988) has adopted the PTI design specifications as UBC Standard 29-4, Part II. However, many feel that this is an overly conservative design for the soil conditions in this region. Structural and geotechnical professional groups in San Diego are currently preparing a design procedure with less stringent requirements to submit to the City Engineer. It is anticipated that an acceptable subsequent design will eventually be accepted for this region.

Other methods used to address expansive soils in Southern California. It is common in Southern California and especially in San Diego County to move large volumes of soil (millions of cubic yards) during the grading of a site. This is most likely due to the large variations in elevation across a typical site. High land values make it profitable to increase the amount of level space during the grading operation by filling in canyons. Because large volumes of soil are typically moved, it is possible to excavate the existing near-surface expansive soil and place it, properly compacted, at a significant depth below the proposed finish grade. This places the soil below the assumed depth of moisture variation.

Another option is to cap the expansive soil with a minimum of 3 feet of nonexpansive soil. This has the same effect as placing the expansive soil at depth. The difference is that the expansive soil is not actually excavated; it is simply covered in place. These methods become economically feasible when the grading of the site requires large amounts of soil to be moved to prepare the building pads. Adding some control over monitoring where nonexpansive or expansive soil will be placed does not affect the cost significantly.

Lime treatment is also becoming more common. The addition of lime reduces the expansion potential of the soil, as previously described in the section on Dallas. This is also cost-effective for the same reason described above. That is, the earth-moving equipment necessary to mix the lime into the soil is generally already present on site for the mass grading.

SAN FRANCISCO BAY AREA

Area Description, and Geologic and Soil Conditions

The San Francisco Bay area is characterized by very hilly terrain surrounding the San Francisco Bay and the Sacramento River Delta. The bay and the extensive river delta were originally formed from the drowning of block-faulted and river-cut valleys (Norris and Webb, 1976). The terrain is comprised mostly of marine sedimentary formations. Many of these marine formations contain significant deposits of expansive clay soils.

The Cw for the region is approximately 16. Similar to the other regions described in this chapter, this is a very low climatic rating, implying very seasonal weather patterns that can lead to wetting and drying cycles. These cycles cause the locally expansive soils to shrink and swell. UBC expansion index tests (UBC 29-2) have been measured as high as 280, and plasticity indexes in the 60s have been observed for some of the local deposits.

Predominant Foundation Systems

Three different foundation systems are commonly used in the San Francisco Bay area to minimize the potential damage caused by expansive soil movement. First, similar to the Denver and Dallas areas, drilled shaft foundation systems are commonly used for lightly loaded structures. Second, post-tensioned slab systems are also employed at sites with expansive soil deposits. Third, a structural slab is used in conjunction with the removal of 3 feet of the expansive soil, which is replaced with non-expansive material. Again, the decision of one system over the other depends on a variety of factors such as: (1) the expansiveness of the soil, (2) the builder's budget, (3) the structural tolerance of the proposed building, and (4) the personal preference of the builder, contractor, or engineer. The presence of a relatively high groundwater table may also affect this decision. Some counties on the north side of the San Francisco Bay area typically have a shallow groundwater table. The common use of drilled shafts for commercial structures and the predominant use of slab-on-grade systems for residential structures is not as prevalent here as it is in Dallas.

Drilled shaft systems in the San Francisco Bay area. As in Denver and Dallas, the design concept for drilled shafts in expansive soils in the San Francisco Bay area is to extend them deep enough to develop enough side friction to offset the potential uplift forces of the expansive soil in the zone of seasonal moisture variation. This zone is typically assumed to be approximately 4 feet in depth for this region. The shafts range in depth from 5 to 10 feet, with the majority of them in the range of 7 to 10 feet. One study revealed that design depths of the shafts have been increasing. In the 1970s, typical depths were 3 to 5 feet; now they are predominantly 7 to 10 feet. It is thought that this is because estimates of the depth of moisture variation have become more conservative over the years. Increased availability of high-speed drill rigs has kept the installation time of deeper shafts relatively short, thus keeping down costs. High real estate values and litigation costs may be one reason why engineers have become more conservative in their estimate of the depth of moisture variation.

The shaft diameters range from 10 to 16 inches and are reinforced to their full length with one to four No. 4 bars, depending on the expansiveness of the soil. The shafts are typically spaced approximately 6 feet apart in an effort to ''pin'' down the structure. The grade beams are typically 6 to 8 inches wide and 16 to 24 inches deep with one No. 4 bar near the top and one No. 4 bar near the bottom. A void is occasionally created beneath the grade beam to allow the soil to expand without contacting the grade beam. This practice is not as widespread as in Denver and Dallas and the void is generally created using methods other than a cardboard void box. A structural wood floor is commonly constructed in conjunction with the drilled shaft and grade beam system. This is probably because more expensive homes are built in this region (the average price of a new home is close to $200,000) and the raised wood floor is perceived as a luxury item.

Post-tensioned slab foundations in the San Francisco Bay Area. Post-tensioned slab construction is becoming more popular in this area for sites with expansive soil. This system is most often used in large subdivisions where a small difference in the cost of a post-tensioned slab versus a drilled shaft system can result in large savings when multiplied by 100 to 200 homes. The main difference in cost is due to the fact that once the post-tensioned slab is poured, the floor system exists. With the drilled shaft system the floor support system must still be constructed, whether it is a floating slab or a raised structural floor.

The problems that have occurred in the San Francisco Bay area with post-tensioned slabs are similar to those in other areas where they have been used. Most of the problems have been related to an unfamiliarity with design and construction procedures. Again, it is recommended that one adhere to the recommendations of the Post-Tensioning Institute manual for slab-on-grade construction or refer to UBC Standard 29-4 (International Conference of Building Officials, 1988).

Removal of 3 feet of expansive material. This option is occasionally used at projects where relatively large volumes of soil will be moved to grade the site. The mass grading enables the geotechnical engineer to specify which soils need to be buried and which materials are acceptable to place within the upper 3 feet of the building pad. Once this has been accomplished, structural slab foundations are used in the event that there is still some movement of the expansive soils at depth.

ADVANTAGES AND DISADVANTAGES OF FOUNDATION SYSTEMS

The above study of four areas in the United States with expansive soil conditions shows there are four general approaches used to reduce the potential damage caused by expansive soil movement beneath a lightly loaded structure. These approaches consist of either a specific type of foundation system, a modification of the subgrade, or a combination thereof. More specifically they are: (1) drilled shaft and grade beam, (2) post-tensioned or structural slab-on-grade, (3) heavily reinforced continuous and isolated footings, and (4) removal and replacement of, or treatment of, the expansive subgrade.

> The least expensive system or the system that "everyone is familiar with" may not be the most effective for a specific site.

The decision as to which approach to adopt depends on a variety of factors including cost and personal preference of the designer. However, the ultimate decision must be based on the individual site conditions. The least expensive system or the system that ''everyone is familiar with'' may not be the most effective for a specific site. With this in mind, advantages and disadvantages of the first three systems are listed in the following pages. Removal and replacement/subgrade treatment is generally used in conjunction with the other three systems to reduce costs and increase their effectiveness.

Drilled Shafts with a Raised Floor

Advantages	Disadvantages or Difficulties
When properly constructed, foundation movement will be minimized or eliminated because a portion of the foundation is socketed or anchored in the soil beneath the zone of seasonal moisture variation.	In general, this is a more expensive foundation system to construct, usually due to the raised floor system.
Isolates grade beams and the floor system from the soil by creating a void between the soil and the grade beams.	Requires strict field inspection to assess the soil conditions on the site.

Advantages	Disadvantages or Difficulties
Easily adapted in the field in case of changed conditions. For example, the shafts can be drilled deeper if necessary.	Difficult to predict the uplift shear values of the swelling soil on the shaft.
Cost-competitive with the post-tensioned slab or structural slab in regions where drilled shafts have become common for lightly loaded construction.	Quality control is essential.
Relatively easy to construct in frozen ground during the winter months.	
Long-term reliability and, therefore, less possibility of requiring future repairs.	

Post-Tensioned and Structural Slab-on-Grade

Advantages	Disadvantages or Difficulties
The reinforcement causes the foundation to act as a rigid unit. Therefore, if subjected to any differential movement, the slab will impart minimal stresses to the structure itself.	An unfamiliarity with proper design and construction procedures exists, especially with post-tensioned systems.
In general, it is less expensive than a drilled shaft system with a raised floor.	Quality control is essential, especially for the post-tensioning system.
Unaffected by shallow groundwater levels during construction.	Heaving of the soil may cause the slab to tilt, requiring remedial work in the future.
Cracking of the slab may be reduced if properly reinforced. Developers are willing to spend the extra money to avoid potential litigation over nonstructural cracks in a slab. Crack control joints are not necessary for post-tensioned design.	Placing the slab directly on the ground causes the moisture in the soil to stop transpiring. Capillary rise will then create a buildup of moisture beneath the slab, which may result in heaving. In general, it is difficult to control the subgrade moisture content before and after slab placement.
	Requires subgrade treatment before slab placement.
	Geotechnical engineers are not accustomed to providing slab design parameters in their reports. Improper

Advantages	Disadvantages or Difficulties
	estimates of edge moisture variation distance may lead to edge curling and racking of the structure.
	Difficult to construct at the basement level.

Continuous and Isolated Spread Footings

Advantages	Disadvantages or Difficulties
In general, this is the least expensive system	As the potential for expansion increases, this system generally is no longer appropriate.
Many builders and contractors are familiar with this system.	Typically cannot generate enough of a load to offset the upward force due to the soil expanding.
Full-time inspection is typically not necessary.	Costs can become high if large amounts of steel, concrete, and labor are necessary to strengthen the footings against expansion.
	Time is required to clean the bottom of the excavation of any loose soil or debris.
	Requires subgrade treatment and or moisture control for the slab.
	Subject to differential heave of the footings and the floor slab, causing racking to the structure and necessitating future repair.

COMMON DESIGN AND CONSTRUCTION OVERSIGHTS

The systems and methods discussed above all have advantages and, if properly implemented, can perform quite well. However, many of the repair dollars that are spent each year are directly related to construction and design oversights. An example of this is a distressed foundation on expansive soil in Vacaville, California, a rural community located approximately 50 miles northeast of San Francisco in the foothills that form the western edge of the Sacramento Valley. The soil conditions in the

Sacramento Valley consist mostly of sandy and silty soils of a nonexpansive nature and are typically stable enough for continuous or isolated spread-footing foundations to be appropriate. Many areas within the Vacaville foothills do contain sites with expansive soils.

The foundation system for this structure consists of drilled shafts with a raised wood floor. Unfortunately, the drilled shafts are only 30 inches deep. The Vacaville region has an active zone approximately 3 to 4 feet deep. Therefore, the full length of the shafts is within the active zone of seasonal moisture variation. Under these conditions the shafts function as round spread footings on 5-foot centers. Presumably, the builder was aware that drilled shafts are commonly used in the San Francisco Bay area for sites with expansive soil. However, because the builder worked on the fringe of the area and had built many more homes on nonexpansive sites, the basic design concept of socketing the shafts in the nonactive zone was overlooked. This oversight has led to cracking of the grade beams, a warped floor, and expensive repairs.

With these types of oversights in mind, some of the more common design and construction errors related to construction at sites with expansive soil are listed below. They are categorized by foundation system and include a brief explanation and/or a suggested method. There are, of course, many other oversights that are not listed here, and readers are encouraged to refer to additional references.

Drilled Shaft Oversights

- **Underloaded shafts.** For expansive soil design, the load on each shaft should be maximized by spacing the shafts as much as possible while maintaining a reasonably sized grade beam. By doing this, the downward load on each shaft will be larger and will thus help to offset the potential uplift force of the expansive soil. Live loads should not be considered in the design, in terms of providing an additional offsetting axial load, because they will not always be present. Underloaded shafts are more common beneath interior columns, where loads typically are not as high. In some regions of the country the standard practice is to construct as many shafts as reasonably possible. The idea is to "pin" down the structure. However, by using this method, the weight of the building on each shaft becomes insignificant and does not add any resistance to the expansive soil.

- **Shafts not drilled deep enough.** The shafts should be deep enough to mobilize enough friction in the nonactive zone to offset the potential uplift forces on the shaft in the active zone. As mentioned above, interior column shafts are typically not loaded as highly. Depending on the soil conditions, consideration should be given to constructing deeper shafts at these locations to mobilize more friction in the nonactive zone of the soil. The length of the shafts can be minimized somewhat by constructing them with shear rings in the nonactive zone, thereby increasing the frictional resistance to uplift. Design parameters are presented by Reese and O'Neill (1988) and by Taylor (1990).
- **Constant length shafts regardless of surface and subsurface conditions.** As mentioned above, it is important to ''socket'' the shafts below the zone of seasonal moisture variation. Some designs may call for a certain amount of embedment into a specific stratum of soil. The design may then consist of constant length shafts across the site based on one or two exploratory borings. It is possible that the depth to the soil stratum may vary across the site, necessitating shafts of different lengths. Another common design oversight related to shaft length has to do with varying surface grade elevations. For example, a home may be constructed with a basement at one grade and a garage at higher grade. In this case it may be necessary to construct longer shafts to support the garage, depending on the subsurface conditions, to ensure proper embedment. It is desirable to have a geotechnical engineer on site during shaft construction to see that the shafts are in competent material and sufficiently within the depth of the nonactive zone.
- **Overpour or ''mushrooming'' of the shafts.** If the concrete is allowed to overflow the drilled shaft a horizontal surface is created, which the expansive soil can push up on. Along the vertical sides of the shafts approximately 15% of the swell pressure is transferred to the shaft (Chen, 1988). However, at a horizontal contact 100% of the pressure is transferred to the concrete because the soil is swelling perpendicular to the surface. Any excess concrete poured outside the shaft should be immediately removed before it cures with the shaft.
- **Honeycombing of concrete.** This condition results from using concrete with too low of a slump. The slump is a measure of the stiffness or workability of the concrete. A low slump indicates a ''stiff'' concrete mix. It is impractical to compact the concrete at depth by

vibration. A high-slump concrete design will compact under its own weight. Low-slump concrete will lead to a poor bond with the soil. A poor bond can result in heaving of the shaft. It can also create a relatively free pathway for water to reach the "inactive" zone, causing it to swell. High-slump concrete is also necessary to ensure that the concrete can flow through a steel reinforcement cage and fill the full extent of a bell (if used) at the base of a shaft.

- **Loose material in the bottom of end bearing shafts.** Loose material in the bottom of shafts designed for end bearing may lead to settlement in the future. It is important to clean out any loose soil carefully at the base of the shaft so the concrete is poured directly on the intended bearing soil layer. It is also strongly recommended that, for safety reasons, the shafts be fully poured on the same day as they are drilled. This also minimizes the probability of soil sloughing or surface material being pushed into the hole. It is a good idea to have the steel and concrete materials and laborers on site while the shafts are being drilled. This way the shafts can be poured with minimal delay. In the event that it is not possible to pour the shaft on the same day that it is drilled, the shaft should be covered with a hole cover and properly marked by the contractor to ensure that no one steps into the shaft. Once the shaft is ready to be poured, it should be checked one final time to make sure that excessive amounts of loose material have not sloughed into the hole. Each site should be evaluated for how much slough is excessive, depending on the nature of the loose material and the sensitivity of the structure.
- **Misplaced shafts.** A shaft in the incorrect location may create a condition where a column is supported by a floor slab-on-grade and the shaft does not have any load on it at all. This condition can lead to the pushing of the shaft out of the ground due to the missing dead load or racking of the structure in the event of floor slab heave. It may also lead to a bearing failure to adjacent shafts that are supporting the load intended for the misplaced shaft.
- **Too much steel for the size of the shaft.** The diameter of the shaft should be considered in the design of the steel. A cage of steel may be very difficult to place in a 10-inch diameter shaft. It may be more practical to increase the size and reduce the number of the reinforcing bars.
- **Water in the shaft.** If there is water in the shaft, the concrete should be pumped or tremied (placed through a pipe) to the bottom

of the shaft to displace the water and to ensure that the concrete mixture does not segregate. The concrete should never be poured directly through a column of water. The *Drilled Shaft Inspector's Manual* (ADSC, 1989) states that tremie placement is necessary if the seepage rate is such that more than 2 inches of water would enter the shaft in the time that it would take to place enough concrete in the hole to balance the maximum head of the groundwater. An adequate pressure head should be maintained during the pour to ensure that any loose soil from the sides of the shaft is not mixed in with the concrete.

- **Void box location.** If the void box is placed at the finish grade level, when it deteriorates an opening is created for surface water to flow under the structure. Another problem that is created is the easy access for rodents, insects, etc., to get beneath the structure. It is recommended that the void box be placed slightly below finish grade and then either backfilled with a compacted, fine-grained, nonexpansive material or sealed with a plastic liner to inhibit the flow of surface water beneath the structure.
- **Void box integrity.** It is important to maintain the integrity of the void box until the grade beam has been poured and cured. If the void box is crushed or becomes wet before the grade beam has been constructed, the design void will not be attained.
- **Presentation of drilled shafts on construction plans.** It is recommended that the shafts be numbered on the set of plans to differentiate among them. This is especially important for designs that call for a variety of depths and reinforcement schedules. Additionally, it is suggested that circles on the plans be reserved only for shaft locations. In the field, any circles on the plans may be construed as a shaft location. An extra shaft in the ground without any designed dead load could lead to large amounts of heave and threaten the integrity of the foundation.

Monolithic Post-Tensioned and Conventionally Reinforced Slab-on-Grade Systems and Other Slab-on-Grade Oversights

- **Inadequate post-tensioned design.** Because the post-tensioned slab-on-grade system is still relatively new, it is often inadequately designed. Designers should insist that geotechnical engineers provide them with the necessary soil information, as outlined in the PTI manuals and accepted by UBC Standard 29-4 (International Conference of Building Officials, 1988), for slab-on-grade design and field procedures.

- **Delaying stressing.** Shrinkage cracks may occur in the slab before the stressing of the strands is permitted. The PTI (1989b) recommends that the concrete reach a compressive strength of 2000 psi before stressing the strands. This may take from 3 to 10 days. The PTI (1989b) also adds that partial stressing is permitted before attaining 2000 psi in order to control shrinkage cracks as long as the stress being applied does not exceed the current strength of the concrete as tested by cylinders.
- **Too much time between subgrade moisture conditioning and slab pour.** It is recommended that the slab be poured as soon as possible after any moisture conditioning of the subgrade has been performed. This time should not be more than two days. If too much time elapses, the subgrade soil will begin to decrease in moisture content and may begin to shrink. Once the slab is placed, the moisture in time will build up beneath it because of moisture migration from the surrounding soil, without the drying effects from the sun. The increase in moisture can cause the soil to swell and stress the slab. Even if the subgrade is not soaked or moisture conditioned, its natural moisture content can decrease if the slab is not poured soon after it has been graded.
- **Interior partitions constructed on floating slabs.** Because floating slabs have a tendency to move as the soil moves, they can cause racking to the structure by moving interior partitions constructed upon them. An alternative design is to hang the partitions from the ceiling and connect them to the floor with a slip joint. The slip joint allows movement of the slab to occur without any stress on the partitions. Figure 4.13 shows typical slip joint configurations for a hung part and a floor-supported partition wall. It is also important to ensure that the sheet rock and/or wall paneling does not rest on the floor slab.
- **Improper wire mesh placement.** Welded wire mesh is often placed incorrectly within the slab. The mesh is the most effective when placed 2 inches from the slab surface (NAHB Research Foundation, 1983). The typical method is to pull the mesh up through the

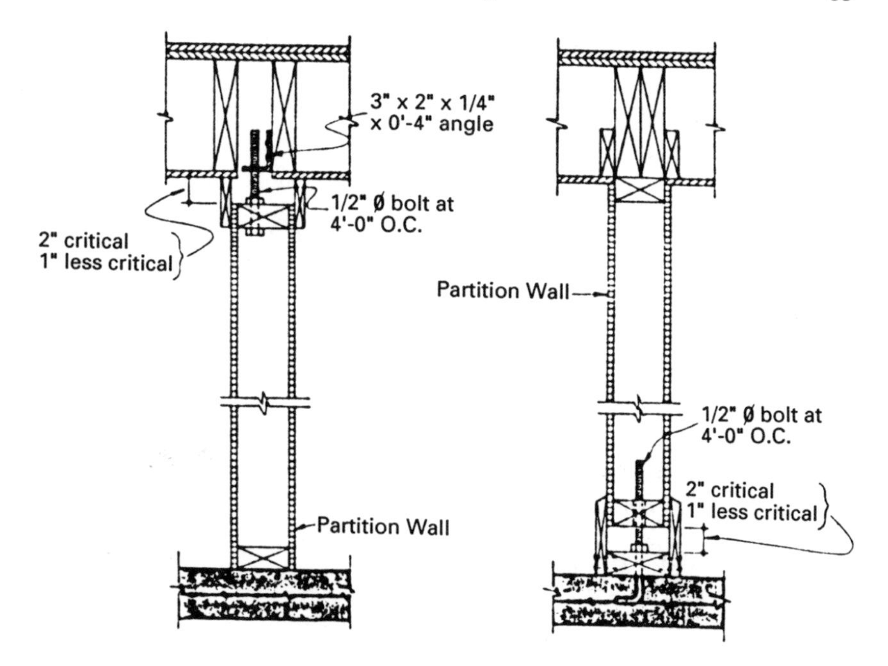

Figure 4.13 Examples of wall slip joints; floor supported and hung partion. (Reprinted by permission of Elsevier Science Publishers, Amsterdam, The Netherlands, from Chen, 1988.)

concrete once it has been poured. This has the disadvantage of pulling soil into the concrete mixture as well as uneven placement. Placing the mesh on blocks or risers is impractical because the flexibility of the mesh would require a large amount of blocks to lift the mesh completely off the ground. The only truly effective way to place the mesh is to pour the slab to the mesh height, place the mesh, and then complete the pour. The disadvantage of this method is higher labor costs.

- **Relying on wire mesh for structural support.** The purpose of wire mesh is to keep concrete shrinkage cracks from widening. It does not eliminate cracking; it controls cracking and helps keep the aggregate in the concrete interlocked to prevent vertical displacement on either side of the crack. Mesh should not be relied on to increase the flexural strength of the slab, especially for sites with highly expansive soil. Reinforcing steel should be used if additional flexural strength is required for possible differential movement of the subgrade.

Continuous and Isolated Spread-Footing Oversights

- **Not appropriate for expansive conditions.** In general, for highly expansive soil conditions, continuous wall and isolated spread footings are not appropriate. This is because the footings cannot generate enough dead load pressure to offset the potential uplift pressures associated with the expansive soil. Even though additional reinforcement may help retain the integrity of the footings, differential heave can cause racking to the structure.
- **Inadequate subgrade treatment.** If continuous footings are going to be used, it is recommended that the subgrade, at the very least, be treated in some way that will enable it to either maintain a constant moisture level or reduce the potential expansiveness of the soil beneath the continuous footings and the floor slab. Examples include: soaking or prewetting the soil, water injection, and lime addition.
- **Relying strictly on empirical data.** Many local areas may rely strictly on empirical data from past experience in the design of footing foundations. This can be dangerous when a site contains highly expansive soil. Extensive laboratory tests should be conducted in this case to evaluate the choice and design of the footing foundation.

General Design and Construction Oversights

- **Inadequate or no subsurface investigation.** Soil is an extremely variable material. It is almost always less expensive to conduct a proper subsurface investigation up front than to perform repair procedures in the future. This is critical with expansive soil deposits because they have a large potential to damage the structure. Recommendations of the soil report should be followed or surpassed. All too often, corners are cut in an attempt to save costs during construction.
- **Basement wall and grade beam backfill.** There are two common oversights concerning basement wall and grade beam backfill.

First, the wall might be backfilled before the concrete has attained adequate strength to support the soil. One recommendation is to wait until the first floor framework is complete. This provides enough time for the concrete to cure, and the framework also provides additional bracing against the backfill. Second, the backfill usually consists of the native soil that was excavated to construct the basement and has been sitting on the job site losing moisture since the excavation. If this soil is of an expansive nature it should be moisture-conditioned before it is used as backfill material. The material should never be used as backfill in a dry condition or moisture conditioned after it has been placed. This will result in high swell pressures for which the basement walls have not necessarily been designed.

- **Flat work connected to building.** Flat work, such as sidewalks, patios, stairs, and planter boxes should not be attached to the building. These extremely light structures should be allowed to "float" with the changes in volume of the soil. If they are attached to the building the potential for cracking is much higher because one end of the flat work is then fixed (assuming the building does not move).
- **Changes in drainage around the foundation.** Many foundation failures related to expansive soil are caused by homeowners changing the drainage path around the foundation, usually for landscaping purposes. The changes can cause water to puddle at the foundation, resulting in heave. It is important to provide an adequate slope away from all sides of the foundation, and the owner must be made aware of the importance of maintaining this slope. The recommended slope for vegetated areas within the first 10 feet of the foundation is 10% (Colorado Geological Survey, 1987) with a minimum slope of 5% to ensure drainage, and a maximum slope of 15% to minimize erosion. It is also recommended that, at a minimum, roof gutters and appropriate downspouts be installed with appropriate downspout extensions outfalling on splashblocks. Ideally, the downspouts should feed directly into a closed pipe, underground system, that leads into the storm drain system. This method of roof runoff containment has the advantage of protection from owner alterations. Failures are known to occur solely because an owner disconnected or moved a drain extension that was in the way. Figure 4.14 shows a diagram of a typical home with aboveground drainage, and, in this case, drilled shaft support. For paved areas a

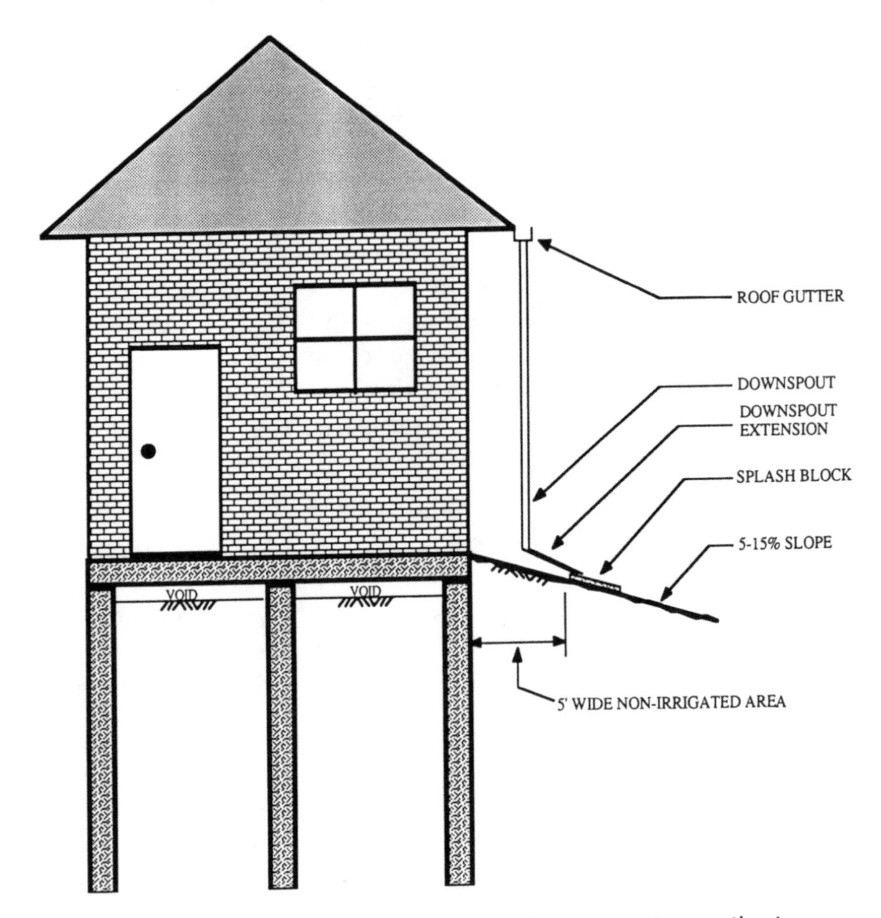

Figure 4.14 Example of drainage for expansive soil site conditions.

minimum slope of 1% may be used; however, a steeper slope is recommended in the event of settlement adjacent to the building. Settlement will result in a flatter slope, which could lead to ponding of surface water.

- **Granular material in utility trenches.** Backfilling utilty trenches with granular material, such as sand, is common practice. For areas with highly expansive material this should be approached with caution because the trenches may act as conduits to transport water to the expansive material beneath the foundation. Related to this is the installation of subsurface drains. Because drains must consist of granular material to be effective, it is often recommended that an impermeable barrier be placed on the foundation side of the drain. In the event of a clog in the drain, the water will then seep into the soil on the side away from the foundation.

- **Vegetation and irrigation near the foundation.** The placement or removal of vegetation near the foundation can significantly alter the moisture content of the soil beneath it. The Colorado Geological Survey (1987) recommends that plant shrubs not be planted within 5 feet of the foundation and trees not be planted within 15 feet. Sprinkler systems should not spray within 5 feet of the foundation. Alternatives to shrubs and trees include native ground cover plants that can help maintain the soil moisture, or perimeter rock gardens. Any watering near the foundation should be done by hand. Readers are encouraged to consult local nurseries for information pertaining to low-water-demanding vegetation and trees that do not have extensive root systems.

RELATED REFERENCES

ADSC: The International Association of Foundation Drilling and DFI: Deep Foundation Institute, *Drilled Shaft Inspector's Manual,* 1st Edition, ADSC, Dallas, 1989.

AMERICAN CONCRETE INSTITUTE, *Guide to Residential Cast-in-Place Concrete Construction,* Report No. ACI 332R-84, American Concrete Institute, Detroit, 1989.

BOYNTON, R. S., and Blacklock, J.R., *Lime Slurry Pressure Injection Bulletin,* Bulletin 331, National Lime Association, Arlington, VA 1985.

BROWN, R. W., *Residential Foundations—Design, Behavior and Repair,* 2nd Edition, Van Nostrand Reinhold, New York, 1984.

BUILDING RESEARCH ADVISORY BOARD (BRAB) OF THE NATIONAL RESEARCH COUNCIL, *Criteria for Selection and Design of Residential Slabs-on-Ground,* Report No. 33 to the Federal Housing Administration, National Academy of Sciences, Washington, DC, 1968.

CHEN, F. H., "Current Status on Legal Aspects of Expansive Soils," *Proceedings of the 6th International Conference on Expansive Soils,* 1987, pp. 353–356.

CHEN, F. H., *Foundations on Expansive Soils,* Elsevier Scientific Publishing Co., New York, 1988.

COLORADO GEOLOGICAL SURVEY, *Home Landscaping and Maintenance on Swelling Soil,* 1st Revision, Special Publication 14, Department of Natural Resources, Denver, 1987.

DONALDSON, G. W., "The Prediction of Differential Movement on Expansive Soils," *Proceedings of the 3rd International Conference on Expansive Soils*, Jerusalem, Academic Press, Vol. 1, 1974, pp. 289–293.

DYWIDAG SYSTEMS INTERNATIONAL, *DYWIDAG Monostrand Posttensioning System*, 1990.

GREER, D. M., and Gardner, W. S., *Construction of Drilled Pier Foundations*, John Wiley & Sons, New York, 1986.

GROMKO, G. J., "Review of Expansive Soils," *Journal of the Geotechnical Engineering Division*, ASCE, Vol. 100, No. GT6, June 1974, pp. 667–687.

HOLTZ, W. G., and Gibbs, H. J., "Engineering Properties of Expansive Clays," *Proceedings*, ASCE, Vol. 80, 1954.

HOLTZ, R. D., and Kovacs, W. D., *An Introduction to Geotechnical Engineering*, Prentice-Hall, Englewood Cliffs, NJ, 1981.

HAUSMANN, MANFRED R., *Engineering Principles of Ground Modification*, McGraw-Hill Book Co., New York, 1990.

INTERNATIONAL CONFERENCE OF BUILDING OFFICIALS, *Uniform Building Code*, ICBO, Whittier, CA, 1988.

JONES, D. E., and Holtz, W. G., "Expansive Soils—The Hidden Disaster," *Civil Engineering*, ASCE, Vol. 43, No. 8, Aug. 1973.

JONES, D. E., and Jones, K. A., "Treating Expansive Soils," *Civil Engineering*, Vol. 57, No. 8, August 1987.

JONES, D. E., and Jones, K. A., "Options for Building on Expansive Clays," *Proceedings of the Up and Down Soil Symposium*, at the University of Colorado, sponsored by the American Society of Civil Engineers, Colorado Section, March 31, 1988.

KENT, H. C., and Porter, K. W., *Colorado Geology*, Rocky Mountain Association of Geologists, Denver, 1980.

KROHN, J. P., and Slossen, J. E., "Assessment of Expansive Soils in the United States," *Proceedings of the 4th International Conference on Expansive Soils*, Denver, 1980, pp. 596–608.

LEE, L. J., and Kocherhans, J. G., "Soil Stabilization by Use of Moisture Barriers," *Proceedings of the 3rd International Conference on Expansive Soils*, Jerusalem, Academic Press, Vol. 1, 1974, pp. 295–300.

LITTON, R. L., and Meyer, K. T., "Stiffened Mats on Expansive Clay," *Journal of the Soil Mechanics and Foundations Division*, ASCE, Vol. 97, No. SM7, July 1971.

MATHEWSON, C. C. ET AL., "Analysis and Modeling of the Performance of Home Foundations on Expansive Soils in Central Texas," *Bulletin of the Association of Engineering Geologists*, Vol. XII, No. 4, 1975, pp. 275–302.

NAHB Research Foundation, Inc., *Residential Concrete,* National Association of Home Builders, Washington, DC, 1983.

Norris, R. M., and Webb, R. W., *Geology of California,* John Wiley & Sons, New York, 1976.

Portland Cement Association, *Joints in Walls Below Ground,* Concrete Report, Portland Cement Association, 1982.

Post-Tensioning Institute, *Design and Construction of Post-Tensioned Slabs-on-Ground,* 1st Edition, Post-Tensioning Institute, Phoenix, AZ, 1989. (a)

Post-Tensioning Institute, *Field Procedures Manual for Unbonded Single Strand Tendons,* Post-Tensioning Institute, Phoenix, AZ, 1989. (b)

Reese, L. C., and O'Neill, M. W., *Criteria for the Design of Axially Loaded Drilled Shafts,* Center for Highway Research, Univ. of Texas, Austin, August 1971.

Reese, L. C., and O'Neill, M. W., *Drilled Shafts: Construction Procedures and Design Methods,* for the U.S. Dept of Transportation, FHWA-HI-88-042, ADSC: The International Association of Foundation Drilling, ADSC-TL-4, Dallas, 1988.

Seed, H. B., Woodward, R. J., and Lundgren, R., "Prediction of Swelling Potential for Compacted Clays," *Journal ASCE, Soil Mechanics and Foundations Division,* Vol. 88, 1962.

Sharp, R. P., *Geology Field Guide to Southern California,* Wm. C. Brown Company Publishers, Dubuque, IA, 1972.

Smith, G. N., *Elements of Soil Mechanics,* 6th Edition, BSP Professional Books, Oxford, 1990.

Taylor, D. L., *Low Capacity Pier Foundations, Denver Metropolitan Area.* Thesis presented to the University of Colorado, Denver, in partial fulfillment of the requirements for the degree of Master of Engineering, 1990.

Texas Highway Department, Test Method Tex-124-E, revised 1972.

Thom, H. C. S., and Vestal, I. B., *Quantities of Monthly Precipitation for Selected Stations in the Contiguous United States,* Environmental Science Services Administration Technical Report EDS 6, 5p., 1968.

Thompson, H. P., "Supplementary Considerations for Slab-on-Grade Design," *Concrete International,* June 1989, pp. 34–39.

Transportation Research Board Committee on Lime and Lime-Fly Ash Stabilization, *Lime Stabilization—Reactions, Properties, Design, and Construction,* State of the Art Report 5, National Research Council, Washington, D.C., 1987.

Wright, P. J., "Lime Slurry Pressure Injection Tames Expansive Clays," *Civil Engineering,* October 1973 and July 1974.

CHAPTER 5

Examples of Construction at Sites with Highly Compressible Clay and/or Organic Soil Deposits

INTRODUCTION

Settlement of buildings has been a concern of engineers and builders for many decades. The theory of consolidation settlement was originally formulated in the early 1920s by Karl Terzaghi. It is generally acknowledged that the formulation of this theory "marked the beginning of modern soil engineering" (Lambe and Whitman, 1969).

In general, the potential for large settlements to occur beneath lightly loaded structures is not as high as it is for buildings with relatively high structural loads. However, for some soil deposits that are highly compressible, the addition of a lightly loaded structure or several feet of fill may cause significant settlement.

The purpose of this chapter is to present examples of approaches to foundation construction (or site alteration) at sites with settlement-sensitive soil deposits in two regions of the country. Examples from the Chicago area and from northeastern New Jersey will be presented. The chapter also discusses some of the advantages and disadvantages of using different foundation systems as well as some of the common construction oversights.

In general, three methods exist to reduce the potential for settlement at sites with highly compressible clay. Two of the methods are also

> For some soil deposits that are highly compressible, the addition of a lightly loaded structure or several feet of fill may cause significant settlement.

applicable to sites with highly compressible organic deposits. Probably the most common method is to remove the compressible soil and either replace it with select fill soil or, if the excavated material is suitable (nonorganic), it may be returned to the excavation site as an engineered fill. Another method is to construct a deep foundation that transfers the weight of the structure to an underlying soil stratum that has more adequate support characteristics. This is typically done with either a driven pile or drilled shaft system. For sites where the majority of the settlement will be caused by the placement of additional fill and not from the structural load, deep foundations are still a viable option. However, there will undoubtedly be some site maintenance related to the placement of more fill once the compressible soil begins to settle, causing the grade around the building to lower. Downdrag forces on the pile or drilled shaft system must also be considered.

The third approach is to preload the soft stratum long before construction begins. Typically this is accomplished by placing fill soil on the site. If the construction schedule is flexible, the fill can be placed a significant period of time before construction begins, thus allowing the compressible soil to settle without placing the proposed structure at risk. The amount of fill that is placed on the site should be equal to, or exceed, the predicted loading on the soil of the proposed structure (and any proposed permanent fill). Although the soft layer may continue to compress for a long time, from a practical standpoint it will become insignificant after a period of time. The amount of time depends on the thickness of the stratum, the underlying and overlying strata, the permeability of the stratum and the stress history of the stratum. This time may range from several months to many years. However, this approach is not applicable to strata with organic constituents, because the organic strata typically do not stop settling; they continue to decay with time.

If the highly compressible stratum is relatively thick, preloading may take longer than the developer can afford to wait. The installation of wick drains is occasionally used to speed the consolidation process. Wick drains are corrugated plastic strips surrounded with filter fabric that are pushed vertically into the ground using a mandrel. The purpose of the wick drains is to shorten the distance that the pore water must travel as it is "squeezed" out of the compressible soil. The shorter path length causes the excess pore pressure, due to the preload, to dissipate much quicker and thus consolidation occurs at a faster rate.

LIGHTLY LOADED CONSTRUCTION AT SITES WITH COMPRESSIBLE CLAY AND/OR ORGANIC STRATA IN CHICAGO, ILLINOIS

Area Description, and Geologic and Soil Conditions

The Chicago area is characterized by relatively flat sweeping terrain. In general, the soil conditions consist of sand and gravel deposits commonly associated with glacial melt. There are also some fairly extensive deposits of interbedded silts and clays commonly referred to as *varved clay deposits*. These deposits are formed by sedimentation at the bottom of lakes at the leading edge of a glacier. During the Pleistocene Epoch, glaciers had advanced as far south as present-day Chicago. The silts were deposited during the summer when the lakes were not frozen. The activity at the surface kept the smaller clay particles in suspension. During the winter months when the glacial lake froze, the water beneath the ice would become calm enough to allow the clay particles to fall out of suspension.

The varved clay deposits are typically fairly soft and consolidate significantly under small loads. Organic deposits, which are common in the area, are also susceptible to large settlements.

Predominant Methods and Foundation Systems

The three commonly used methods of minimizing the potential for settlement at sites with highly compressible clays and/or organic deposits in the Chicago area are: (1) to remove and replace the stratum, (2) to utilize drilled shaft foundations, and (3) to utilize driven pile foundations. The decision as to which method to choose depends on the specific site conditions and cost. These factors are somewhat different from the factors involved in choosing a foundation system for a site with expansive soil, as discussed in Chapter 4. The type of foundation chosen for a site with expansive soil depends on the personal preference of the owner/builder as well as the cost and site conditions. The authors believe this is due to the better overall understanding of the settlement phenomenon. Because the magnitude of the settlement can be more accurately predicted, each of the above methods can be effectively designed to minimize the settlement of the foundation. Therefore, the deciding factor becomes cost, not the performance of the foundation.

Remove and replace. This method is generally used if the highly compressible layer will be within 6 to 7 feet of the final design grade. It is the most common solution for lightly loaded structures because it is less costly than utilizing a deep foundation, and the loads generally are not large enough significantly to affect the soil at depth. If the loads on the floor slab are not large enough to cause significant settlement or if a raised floor slab is going to be constructed, then only the soil in the general area of the footings will typically be removed with a backhoe. The removed soil is typically replaced with crushed concrete or rock. Occasionally concrete is poured to reach the grade of the bottom of the proposed foundation. These materials have the advantage of requiring little or no compaction upon placement. Figure 5.1 is an example of a trench

Figure 5.1 Typical removal and concrete replacement in Chicago.

that has been over-excavated and replaced with concrete. The darker soil near the bottom of the photograph was determined to be highly compressible and thus was removed. The replacement concrete can be seen behind the sheets of plywood.

Once grade is established, the foundation will typically consist of a conventionally formed footing or what is locally known as a *trench footing*. The trench footing is simply a continuous wall footing, which is the same width as the bearing wall. The advantage is that this footing does not need any formwork because the concrete is poured right up to the sides of the trench.

Removal and replacement is limited to the upper 6 to 7 feet primarily for economic reasons. If the compressible stratum is deeper than 6 to 7 feet, it becomes more economical to construct a drilled shaft or driven pile foundation.

Readers are reminded that any excavation deeper than 5 feet falls under the strict guidelines of the Federal Department of Occupational Safety and Health Administration (OSHA) as related to the allowable steepness of the sideslopes and/or the required shoring. Safety measures are extremely important when excavating trenches or drilling shafts. Excavations that appear sturdy may be deceptive. No life should be placed at risk simply to save time and money!

Drilled shaft foundations in Chicago. Drilled shaft systems are typically used if the compressible layer extends from the near surface to depths greater than 7 feet but less than 30 feet. The shafts are end bearing and transfer the building load to a stratum at a depth that will be able to support the structure without any significant settlement. The shafts are typically placed at each of the building corners. The spacing of additional shafts between the corner shafts is designed to maintain a realistically sized shaft and grade beam to support the column loads. This is typically accomplished with 18- or 24-inch-diameter shafts. If a larger shaft is needed to reduce the bearing load on the soil, it becomes more economical to construct a bell at the base of the shaft to reduce the amount of concrete required to fill the shaft.

Many contractors and engineers prefer 30-inch shafts over 24- or 18-inch shafts regardless of the bearing capacity of the underlying stratum. The larger shaft increases the ease of construction and workability of the shaft, particularly if reinforcing steel is required. If absolutely necessary, it also enables a person to enter the shaft to check the bottom for cleanliness. Safety should be the number-one concern when a person is sent down a shaft. All OSHA regulations should be strictly adhered to. The *Drilled Shaft Inspector's Manual* (ADSC, 1989) also contains a list of suggested safety precautions that should be consulted when sending *anyone* down a drilled shaft.

The shafts can be either belled or straight depending on the strength and cohesiveness of the bearing stratum. If the bearing stratum does not have a significant amount of fines, is not cemented, or, if it is below the groundwater table, it will not be possible to excavate a bell without it collapsing. Bell diameters should not exceed three times the shaft diameter.

At a minimum, there will generally be reinforcing steel placed at the top of the shafts to tie them into the grade beams. The grade beams or basement walls are typically reinforced with a minimum of one bar in the top and the bottom to span the space between the drilled shafts without relying on any support from the soil itself. The floor system generally will consist of a raised wood floor for the structure and a slab-on-grade for the garage.

There are some site conditions that may increase the difficulty and the cost of properly constructing a drilled shaft foundation. Examples of these conditions include: (1) a soft compressible stratum in excess of 30 feet thick at or near the surface, (2) sites with soils that may collapse or slough during drilling of a shaft, or (3) sites with a relatively shallow water table above the design depth of a proposed drilled shaft. Larger drill rigs are typically needed for holes greater than 30 feet deep, and drilling rates slow down at greater depths. Sloughing deposits would require the shaft to be cased during drilling, and a high groundwater table would require the concrete to be tremied. Tremied concrete consists of placing concrete below water level through a pipe, the lower end of which is kept immersed in fresh concrete so that the rising concrete from the bottom displaces the water without washing out the cement content. As the shaft fills with concrete the pipe is pulled up within the shaft. The influx of the concrete forces the water out of the shaft. The concrete cannot be poured directly through the column of water because the mix would segregate as it sunk to the bottom of the shaft, losing its strength. None of these conditions eliminates the potential for using drilled shafts; however, increased costs due to these conditions may render piling to be a more economical solution.

Driven pile foundations in Chicago. In general, a driven pile foundation is a more expensive system compared to removal and replacement of the compressible stratum or the construction of a drilled shaft foundation system. One reason for this is that the mobilization of a pile-driving rig may take up to an entire day. If the job is relatively small, the mobilization costs may make up a large percentage of the costs. There are some site conditions, however, for which driving piles is an economical solution. As described above, examples of these conditions include: (1) a soft compressible stratum in excess of 30 feet thick at or near the surface, (2) sites with soils that may collapse or slough during drilling of a shaft, or (3) sites with a relatively shallow water table above the design depth of a proposed deep foundation. When one of these conditions is encountered in the Chicago area, consideration is given to constructing a pile foundation system.

However, because of the higher cost of driven piling, it is rare to use them on residential structures. They are used for lightly loaded commercial structures when the soil conditions warrant it, as discussed above. Another disadvantage of driving piles is the large vibrations that are caused from the driving. The vibrations can cause damage to nearby buildings.

There are two types of driven piling that are used in the Chicago area for lightly loaded construction: (1) timber and (2) cased.[1] The timber piles are typically constructed from 20-ton capacity lumber and are generally used up to a maximum length of 50 feet. This is because that is the maximum length at which the timber piles are generally available.

If the bearing stratum is greater than 50 feet below the existing grade, steel-cased piles will occasionally be used. A cased pile is constructed by driving a steel casing into the ground with a solid mandrel. The mandrel is then withdrawn and the casing is filled with concrete. A hole is often predrilled for a portion of the pile depth in order to facilitate the driving of the steel pile casing.

Other Foundation Systems in Chicago

Post-tensioned slabs have been used in Chicago for sites that are compressible enough to rule out footings, yet are not soft enough to warrant drilled shafts or driven piles. The design is similar to that for expansive soil (see Chapter 4, "Monolithic Reinforced Slab-on-Grade Foundation Systems in Dallas"). The design methodology is to construct a slab that is stiff enough to maintain its integrity based on an allowable

[1]Cased piles are often referred to as pipe piles or thin-walled steel piles.

differential deflection. In general, the post-tensioned system has been successful when it has been used in the Chicago area. It is not commonly used, however, because of an affinity for basement construction in the region. The post-tensioned, slab-on-grade system is difficult to construct at the basement level.

Other options that have been used include: (1)Preloading the compressible clay for a period of time to induce the settlement, (2) mixing lime with the clay and compacting it to increase its strength, or (3) using dynamic compaction (dropping a large weight on the surface) to compact the compressible layer at depth. All of these options have been employed with various degrees of success depending on the site conditions; however, it is not common to use any of these methods in this area for lightly loaded structures because they are usually cost-effective only at large sites.

LIGHTLY LOADED CONSTRUCTION AT SITES WITH COMPRESSIBLE CLAY AND/OR ORGANIC STRATA IN NORTHEASTERN NEW JERSEY

Area Description, and Geologic and Soil Conditions

Soil conditions in northeastern New Jersey are similar to Chicago in that there are some regions with soft, compressible varved clays formed during the Pleistocene Ice Age. The presence of organic deposits is even more prevalent. These deposits are typically associated with the New Jersey Meadowlands, a large marshy area in the northern part of the state. A relatively shallow groundwater table is present in this region of the state.

Predominant Methods and Foundation Systems

There are two common methods of minimizing the potential for settlement at sites with highly compressible clays and/or organic deposits in the northeastern New Jersey area. They are: (1) remove and replace the stratum, and (2) construct driven pile foundations. Similar to Chicago, the decision is dependent on the cost of the method and the site conditions.

Remove and replace. This method is generally used if the highly compressible varved clays, silts, and marsh deposits are located within the upper 10 feet of the existing ground surface. Similar to Chicago, this is the most common solution for lightly loaded structures because it is less costly than utilizing a deep foundation, and the loads are generally not large enough to affect the soil at depth significantly. The compressible soil in the area of the proposed footings is excavated and backfilled with either controlled fill, concrete, or crushed stone. Once backfilled, a typical continuous wall footing or isolated spread footings may then be utilized.

The depth of 10 feet compared with 6 to 7 feet in Chicago is most likely based on the fact that compared to the Chicago area, drilled shafts are rarely utilized in northeastern New Jersey. The remaining option is to use driven piles. As mentioned previously, pile driving can be very expensive. Therefore, in many instances it is more economical to remove up to 10 feet of soil rather than use driven piling. Drilled shafts are not commonly used in this region because of the shallow level of the groundwater table in the areas that have compressible soil deposits.

Driven pile foundations in northeastern New Jersey. The piles used in northeastern New Jersey are commonly designed as friction piles. That is, there is not always a bearing stratum into which they are driven for support. Instead, the capacity of the piles is achieved solely from friction along the sides of the pile. Of course, if there is a bearing stratum within a reasonable depth, the pile will be driven to this depth and will derive its capacity from both friction and end bearing.

Both timber and cased piles are used in this region. Timber piles are commonly 7 to 8 inches in diameter at the tip and are slightly tapered. Generally, the timber piles will be from 15 to 40 feet in length. If a deeper pile is required, a cased pile will be used. In this instance an 8- to 10-inch hollow, thin-walled steel pipe (capped at the driving end) is driven into the ground and then filled with concrete.

As noted in the section on the Chicago area, a disadvantage of driving piles is the resulting large vibrations that can cause damage to nearby buildings.

Other Methods Utilized in Northeastern New Jersey

Methods such as preloading and the installation of stone columns are occasionally utilized to improve the site such that spread and continuous footing foundations become appropriate. Preloading can be used when a compressible layer is at depth or too thick to excavate economically. Stone columns consist of columns of gravel that are placed in the soil using a vibrating device known as a *vibroflot*. The vibroflot is equipped with a water jet that creates a hole through the soft clay. The gravel is added to the hole, falling through the water from the vibroflot. The vibration of the vibroflot as it is removed from the hole compacts the gravel. The installation of the stone columns adds additional support characteristics to the site and thus helps to reduce the potential amount of settlement (Sowers, 1979).

ADVANTAGES AND DISADVANTAGES OF FOUNDATION SYSTEMS

The above study of two geographic regions in the United States shows that there are three general approaches when attempting to reduce the potential for settlement at sites with highly compressible and/or organic soils. Before discussing the advantages and disadvantages of these methods, it should be reiterated that neither a structure (regardless of its weight) nor fill soils should ever be constructed or placed in such a manner that they will derive long-term support from an organic deposit of soil. This condition will always lead to large settlements as the organic material continues to decay.

The three approaches are: (1) removal and replacement of the soft subgrade soils, (2) drilled shaft foundations, and (3) driven pile foundations. The decision as to which method will be used is largely based on economic considerations; however, the individual site conditions typically dictate which foundation system will provide adequate support at the least expensive cost. What follows are some general advantages and disadvantages to keep in mind when assessing the potential use of each system.

Remove and Replace Method

Advantages	Disadvantages or Difficulties
Often it is the least expensive feasible system or method for sites with near-surface compressible strata less than 10 feet thick.	May not be appropriate or economically feasible for depths greater than 10 feet.

Advantages	Disadvantages or Difficulties
Bearing stratum can be visually and physically observed once the unacceptable material has been removed.	Settlement of the replacement material can still occur.
Allows the use of isolated and continuous spread-footing foundations.	Disposal of the removed material may be expensive. It is generally not suitable as backfill at another site, and landscaping firms do not want it because of an overabundance. Transporting it to a sanitary landfill is an expensive option as well as an inefficient use of a landfill site.

Drilled Shafts Method

Advantages	Disadvantages or Difficulties
When properly constructed, foundation movement will be minimized because the shaft will be bearing on a relatively incompressible layer.	Requires strict field inspection to assess the soil conditions at the design depth.
Isolates the grade beam and floor system from the compressible soil by deriving support from an appropriate underlying bearing stratum.	Shafts may settle excessively or fail because of overstressing, due to downdrag forces from adjacent soil settling (see ''Driven Piles'' Subsection in ''Common Design and Construction Oversights'' section in this chapter).
Easily adapted in the field in the event of changed conditions. For example, the shafts can be drilled deeper if necessary.	Quality control is essential, especially at sites where concrete is placed under water. The concrete must be pumped to the bottom of the shaft (tremied) to displace the water from the shaft without segregating the aggregate in the concrete mix.
Relatively easy to construct in frozen ground during the winter months.	
Physical and visual information can be obtained on the bearing stratum once it is exposed in the shaft.	

Driven Piles Method

Advantages	Disadvantages or Difficulties
When point bearing piles are properly installed, foundation movement will be minimized because the pile will be bearing on a relatively incompressible layer.	In general, this is a more expensive foundation system owing to the high mobilization cost of a pile-driving rig.
Isolates the grade beam and floor system from the compressible soil by transferring the load to an appropriate bearing stratum below.	Requires strict field observation to monitor driving depths and penetration rates.
Relatively easy to construct in frozen ground during the winter months	Difficult to assess conditions of the bearing stratum. Unable to inspect the bearing stratum either visually or physically.
Relatively unaffected by groundwater.	Difficult to ensure that the pile has not deflected during driving.
	Piles may settle or fail because of overstressing due to downdrag forces from adjacent soil settling (see ''Driven Piles'' subsection in ''Common Design and Construction Oversights'' section in this chapter).
	Large vibrations and noise pollution are created during construction. The vibrations can be forceful enough to cause settlement and/or damage to nearby buildings.

COMMON DESIGN AND CONSTRUCTION OVERSIGHTS

All systems and methods discussed above have advantages, and if they are implemented properly they can perform quite well. However, many of the repair dollars spent each year are directly related to construction and design oversights. Below are some of the more common oversights related to construction at sites with soft compressible soil. Following each oversight is a brief explanation and/or suggested method. These oversights are not necessarily related exclusively to the regions discussed in this chapter; they are general oversights related to lightly loaded construction at sites with compressible clays and/or organic deposits. There may be other construction and design oversights that are not listed here, and readers are encouraged to consult additional references pertaining to their site and foundation system.

Remove and Replace Oversights

- **Undercut is not areally extensive enough.** The load of a foundation spreads out wider than the foundation itself. Therefore it is necessary to remove some of the soil beyond the limits of the building footprint to ensure that all potentially compressible material is removed from the influence of the proposed footings. An example of this situation is depicted in Figure 5.2. The areal extent of the influence of a footing is commonly approximated with depth as sloping at a ratio of 2:1 (vertical to horizontal) as shown in Figure 5.2 (French, 1989; Das, 1984). The pressure exerted on the soil is dissipated with depth. The pressure exerted on the soil at depths greater than four times the width of a continuous wall footing generally becomes insignificant (French, 1989). It can be seen in Figure 5.2 that trenching only beneath the proposed footings would

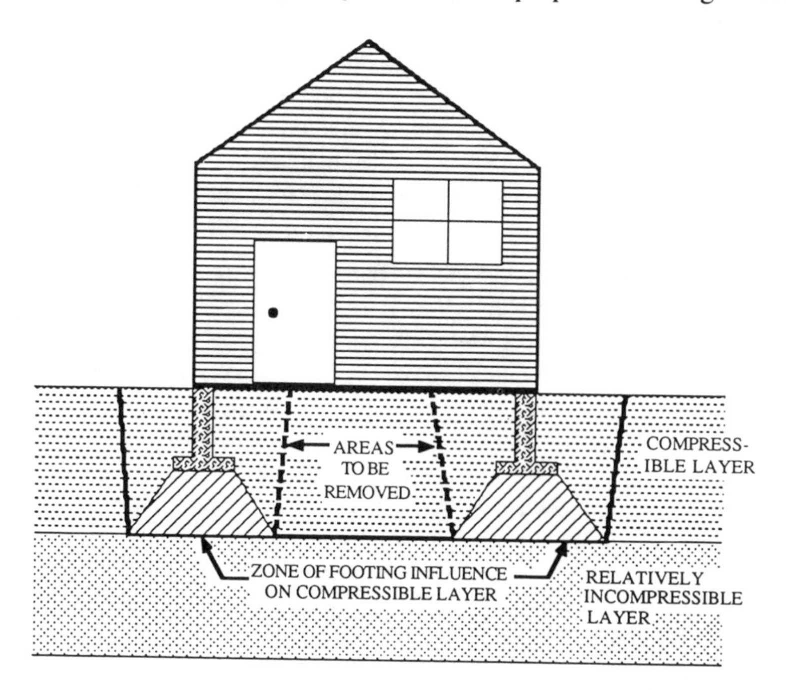

Figure 5.2 Example of areas of removal of compressible soil beneath footings.

leave compressible material in place that would be influenced by the footing load. This could lead to excessive settlement of the footings. Depending on the extent of the project and whether there will be a raised floor or a slab-on-grade, it may be more cost-efficient to over-excavate beneath the entire building location as shown by the solid dark line in Figure 5.2 or to excavate only in the vicinity of the footing locations as shown between the dashed and solid dark lines. It is also important to keep in mind the effects of any additional fill being placed during site grading. If a significant amount of fill will be added to the site to attain finish grade, it may be necessary to over-excavate the entire site and replace it with select fill to ensure that there will not be any excessive settlement.

- **Failing to remove organic soils.** Foundations should not be constructed in such a manner that they will transfer any load to organic soils. This oversight commonly occurs in practice when an adequate soil investigation was not performed and the organic soil went undetected. This oversight includes the placement of compacted fills on top of organic soil deposits. The fill may be properly compacted, but the weight of the fill can cause the organic soil to settle.
- **Assuming that the backfill will not settle.** All backfill material will settle to some degree. These settlements will be significantly less than the settlements that would have occurred in the soft compressible material. Nevertheless, they must still be incorporated into the foundation design. As an example, Monahan (1986) indicates that a *dumped* granular backfill may settle as much as 10% of its original thickness.
- **Loose material in the excavation.** It is important that any loose material that may have fallen or sloughed into the excavation is either cleaned out or properly compacted. It is also common for a contractor to excavate deeper than necessary. Once this is recognized, the excavated material can be dumped back into the trench. This can lead to large settlements if it is not compacted properly. Ideally, the over-excavated material should remain out of the trench and additional concrete, select fill, or the intended backfill material should be used.

Driven Pile Oversights

- **Underestimation of downdrag forces.** Often several feet of fill soils may be placed on a site for a variety of reasons, including creating positive drainage from the building pad, or raising a site above the 100-year flood level. The fill may cause the compressible soil layers to settle. As they settle, friction forces will act downward on any piles that pass through the settling stratum. The downdrag forces may be large enough to cause the piles to settle excessively or fail due to overstressing. These forces must be considered, even for sites without imported fill, because the compressible stratum may still be settling under its own weight or the weight of overlying strata.

- **Damaging piles due to overdriving.** Some designs may call for a specific length of penetration of the pile into the ground. These depths may be based on information from one or two soil borings. As the pile-driving progresses, an area of the site may contain a dense or stiff layer that was not detected in the subsurface investigation. Because the design called for a minimum depth of penetration, some of the piles may be damaged in an attempt to penetrate this layer. The ultimate driving capacity of the piles in terms of blows per unit length of penetration should be known by field personnel. The driving rate should be monitored and driving should be terminated if the driving capacity has been reached. Fleming, Weltman, Randolph et al. (1985) shows a number of examples of piles that were damaged from overdriving. Designs should be flexible enough to allow for change in the event of unanticipated subsurface conditions.
- **Vibrations.** Pile driving creates a large amount of ground vibration. Consideration must be given to the potential for damage to neighboring buildings. Damage can occur if the vibrations densify the soil beneath an adjacent building, causing it to settle. Fleming, Weltman, Randolph et al. (1985) suggest preboring the ground where the pile is to be driven so as to reduce the vibrations.
- **Presentation of driven piles on construction plans.** It is recommended that the piles be numbered on the set of construction plans so as to differentiate among them. This is especially important for designs that call for a variety of pile depths and capacities. Additionally, it is suggested that circles on the plans be reserved only for pile locations.

Drilled Shaft Oversights

- **Loose material in the bottom of end bearing shafts.** Loose material in the bottom of shafts designed for end bearing may lead to settlement in the future. It is important to carefully clean out any loose soil at the base of the shaft so the concrete is poured directly on the intended bearing soil layer. It is also strongly recommended that, for safety reasons, the shafts be fully poured on the same day as they are drilled. It also minimizes the probability of soil sloughing or surface material being pushed into the hole. It is a good idea to have the steel and concrete materials and laborers on site while the shafts are being drilled. This way the shafts can be poured with a minimal delay. In the extreme case that it is not possible to pour the shaft on the same day that it is drilled, the shaft should be covered with a hole cover and properly marked by the contractor to ensure that no one steps into the shaft. Once the shaft is ready to be poured, it should be checked one final time to make sure that excessive amounts of loose material have not sloughed into the hole. Each site should be evaluated for how much slough is excessive, depending on the nature of the loose material and the sensitivity of the structure.

- **Constant length shafts regardless of surface and subsurface conditions.** Some designs may call for a certain amount of embedment into a specific stratum of soil. The design may consist of constant-length shafts across the site based on one or two exploratory borings. It is possible that the depth to the soil stratum may vary across the site, necessitating different-length shafts. Another common design oversight related to shaft length has to do with varying surface grade elevations. For example, a home may be constructed with a basement at one grade and the garage at a higher grade. In this case it may be necessary to construct longer shafts to support the garage, depending on the subsurface conditions, to ensure proper embedment. It is generally recommended that a geotechnical engineer be present on site during shaft construction to see that the shafts have been drilled into competent material.
- **Honeycombing of concrete.** This condition results from using concrete with too low of a slump. The slump is a measure of the stiffness or workability of the concrete. A low slump indicates a "stiff" concrete mix. It is impractical to compact the concrete at depth by vibration. A high-slump concrete design will compact under its own weight. Low-slump concrete will lead to a poor bond with the soil. High-slump concrete is also necessary to ensure that the concrete can flow through a steel reinforcement cage and fill the full extent of a bell at the base of a shaft.

- **Water in the shaft.** If there is water in the shaft, the concrete should be pumped or tremied to the bottom of the shaft to displace the water and to ensure that the concrete mixture does not segregate. The concrete should never be poured directly through a column of water. The *Drilled Shaft Inspector's Manual* (ADSC, 1989) states that tremie placement is necessary if the seepage rate is such that more than 2 inches of water would enter the shaft in the time that it would take to place enough concrete in the hole to balance the maximum head of the groundwater. An adequate pressure head should be maintained during the pour to ensure that any loose soil from the sides of the shaft is not mixed in with the concrete.
- **Presentation of drilled shafts on construction plans.** It is recommended that shafts be numbered on the set of construction plans to differentiate among them. This is especially important for designs that call for a variety of depths and reinforcement schedules. Additionally, it is suggested that circles on the plans be reserved only for shaft locations. In the field, any circles on the plans could be construed as a shaft location.

General Design and Construction Oversights

- **Inadequate or absence of a subsurface investigation.** Soil is an extremely variable material. It is always less expensive to conduct a proper subsurface investigation up front than to perform repair procedures later. This is critical for compressible soil deposits because they have a large potential to damage the structure. Generally, foundations can be designed or sites can be improved to offset the potential damage due to the settlement of a compressible soil deposit. The majority of foundation failures due to settlement are related to an unknown condition resulting from an inadequate subsurface investigation or neglecting the recommendations of a soil report. The recommendations of the soil report should be strictly followed. Too often corners are cut, especially in residential construction, in an attempt to save money.
- **Dessication.** Mature trees planted close to the foundation can lower soil moisture levels as their roots grow beneath the foundation. If the soil is clayey the reduction of moisture may cause shrinkage of the soil, causing the foundation to settle. The reverse process can occur if a mature tree is removed from a location near the

Soil is an extremely variable material. It is almost always less expensive to conduct a proper subsurface investigation up front than to perform repair procedures later.

foundation. The moisture level of the subgrade soil will rise due to capillary action and potentially cause some heave. The construction of a basement or a deep foundation system such as drilled shafts or driven piles will avoid this potential problem because the tree roots generally do not have an impact that deep into the soil. Another rule of thumb is to plant trees a minimum horizontal distance from the foundation equal to the ultimate height of the tree.

RELATED REFERENCES

ADSC: The International Association of Foundation Drilling and DFI: Deep Foundation Institute, *Drilled Shaft Inspector's Manual*, 1st Edition, ADSC, Dallas, 1989.

American Concrete Institute, *Guide to Residential Cast-in-Place Concrete Construction*, Report No. ACI 332R-84, American Concrete Institute, Detroit, 1989.

Bowles, J. E., *Engineering Properties of Soils and Their Measurement*, 3rd Edition, McGraw-Hill Book Co., New York, 1986.

Brown, R. W., *Residential Foundations—Design, Behavior and Repair*, 2nd Edition, Van Nostrand Reinhold, New York, 1984.

Building Research Advisory Board of the National Research Council, *Criteria for Selection and Design of Residential Slabs-on-Ground*, Report No. 33 to the Federal Housing Administration, National Academy of Sciences, Washington, DC, 1968.

Das, B. M., *Principles of Foundation Engineering*, PWS Publishers, Boston, 1984.

DYWIDAG Systems International, *DYWIDAG Monostrand Posttensioning System*, 1990.

Fleming, W. G. K., Weltman, A. J., Randolph, M. F. et al., *Piling Engineering*, John Wiley & Sons, New York, 1985.

French, S. E., *Introduction to Soil Mechanics and Shallow Foundations Design*, Prentice-Hall, Englewood Cliffs, NJ, 1989.

GREER, D. M., AND GARDNER, W. S., *Construction of Drilled Pier Foundations,* John Wiley & Sons, New York, 1986.

HAUSMANN, MANFRED R., *Engineering Principles of Ground Modification,* McGraw-Hill Book Co., New York, 1990.

HOLTZ, R. D., AND KOVACS, W. D., *An Introduction to Geotechnical Engineering,* Prentice-Hall, Englewood Cliffs, NJ, 1981.

LAMBE, T. W., AND WHITMAN, R. V., *Soil Mechanics,* John Wiley & Sons, New York, 1969.

MONAHAN, E. J., *Construction Of and On Compacted Fills,* John Wiley & Sons, New York, 1986.

NAHB RESEARCH FOUNDATION, INC., *Residential Concrete,* National Association of Home Builders, Washington, DC, 1983.

POST-TENSIONING INSTITUTE, *Design and Construction of Post-Tensioned Slabs-on-Ground,* 1st Edition, Post-Tensioning Institute, Phoenix, AZ, 1989. (a)

POST-TENSIONING INSTITUTE, *Field Procedures Manual for Unbonded Single Strand Tendons,* Post-Tensioning Institute, Phoenix, AZ, 1989. (b)

REESE, L. C., AND O'NEILL, M. W., *Criteria for the Design of Axially Loaded Drilled Shafts,* Center for Highway Research, Univ. of Texas, Austin, August 1971.

REESE, L. C., AND O'NEILL, M. W., *Drilled Shafts: Construction Procedures and Design Methods,* for the U.S. Dept. of Transportation, FHWA-HI-88-042, ADSC: The International Association of Foundation Drilling, ADSC-TL-4, Dallas, 1988.

SANDERS, G. A., *Light Building Construction,* Reston Publishing Co., Reston, VA, 1985.

SMITH, G. N., *Elements of Soil Mechanics*, 6th Edition, BSP Professional Books, Oxford, 1990.

SOWERS, G. F., *Introductory Soil Mechanics and Foundations; Geotechnical Engineering*, 4th Edition, Macmillan Publishing Co., New York, 1979.

TERZAGHI, K., *Theoretical Soil Mechanics,* John Wiley & Sons, New York, 1943.

THOMPSON, H. P., "Supplementary Considerations for Slab-on-Grade Design," *Concrete International*, June 1989, pp. 34–39.

TOMLINSON, M. J., *Pile Design and Construction Practice*, 3rd Edition, Palladian Publications Limited, London, 1987.

CHAPTER 6

Examples of Construction at Sites with Uncontrolled Fill and/or Controlled Deep Fills, Differential Fills, and Cut/Fill Transitions

INTRODUCTION

The importance of this chapter is illustrated by E. J. Monahan's observation: ". . . almost no significant engineered construction occurs without the movement of soil from one place to another" (1986). It can be inferred, therefore, that the vast majority of projects will include one or more of the following site conditions: (1) uncontrolled fills, (2) controlled deep fills, (3) differential fills, (4) cut/fill transitions. These different site conditions may each result in differential settlement of the soil. Even lightly loaded structures must be designed for a certain amount of settlement under these conditions. Each of these site conditions is briefly defined below and illustrated in Figure 6.1:

> ". . . almost no significant engineered construction occurs without the movement of soil from one place to another" (Monahan, 1986)

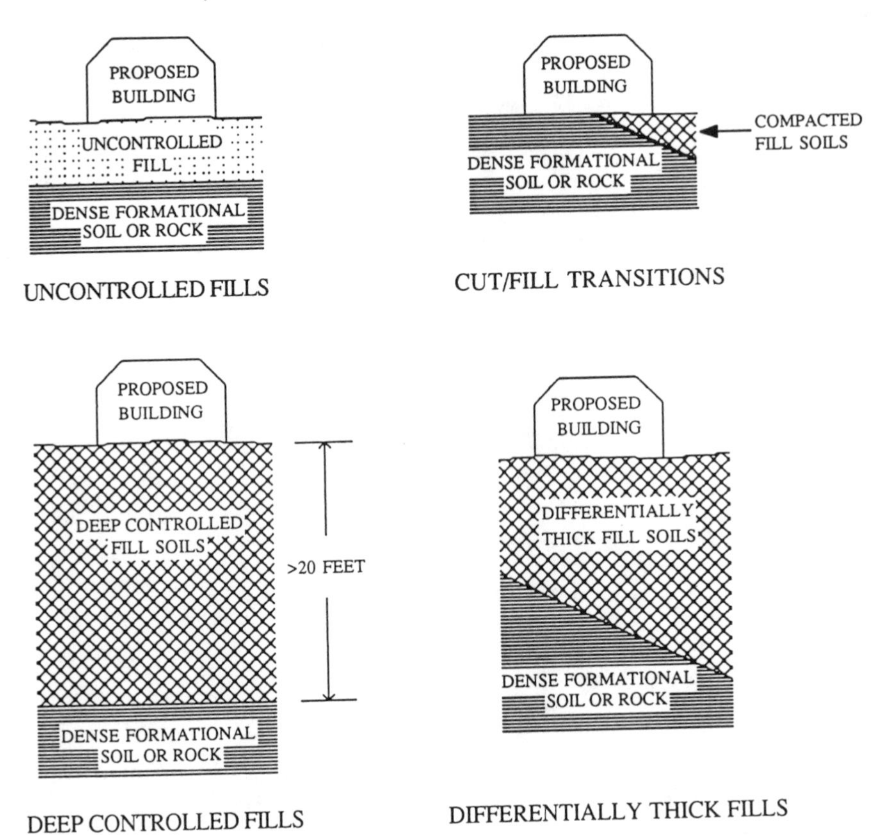

Figure 6.1 Typical fill geometries encountered.

Uncontrolled Fills

This site condition includes fills that were not documented with compaction testing as they were placed; these include dumped fills, fills dumped under water, hydraulically placed fills, and fills that may have been compacted yet there is no documentation of testing or the amount of effort that was used to perform the compaction. These conditions typically exist at sites that may have been graded many years ago before the importance of proper compaction was understood to the extent that it is today.

Controlled Deep Fills

Controlled deep fills are conditions where the soil placement was engineered; however, the thickness of the fill is greater than approximately 20 feet. When fill placements exceed this thickness, settlement of the fill may occur from its own weight, even though it has been controlled and monitored upon placement. Additionally, deep fills may undergo significant settlements if the water content of the deeper portions of the fill increases (Lawton, Fragaszi, and Hardcastle, 1989). The source of the water is typically from landscape irrigation or from pervious strata in canyon walls.

Differential Fills

This site condition can exist at a site that receives different depths of fill across the building pad. As discussed in the above definition, fills are subject to settlement from their own weight. If the fill depth varies across a site it is very likely that this settlement will not be uniform. When the difference in fill depths approach 10 feet or more across a small building pad (e.g., for a residential structure) the potential for damaging settlement becomes a concern. Buildings with more square footage may be able to be supported adequately on greater differential fills because the ratio of the span of the building relative to the differential fill thickness decreases as the building dimensions increase. Each individual case should be examined with the amount of differential fill thickness and subsequent settlement in mind.

Cut/Fill Transitions

A cut/fill transition is essentially a special case of a differential fill condition. The difference is that the entire site does not receive fill soil. A portion of the building pad may have soil removed—a process known as cut—while another portion may receive fill soil. This condition commonly occurs at steeply sloping sites that have been made into a level building pad. The structure is subject to the same differential settlement as that described above for a differential fill. If the cut was into nonexpansive, relatively incompressible and formational material, then very little settlement would be expected at this portion of the lot for a lightly loaded structure. In some cases the cut portion of the lot may rebound because of the unloading of the soil. Conversely, the fill portion could settle under its own weight and distress the structure.

The following sections discuss foundation solutions that are commonly used to minimize potential foundation settlement or distress for site conditions as discussed above. Regional examples are given for San Diego, California, and Miami, Florida. The end of the chapter includes common construction and design oversights regarding both the placement of fill soil and the approach to the fill geometries presented above.

LIGHTLY LOADED CONSTRUCTION AT SITES WITH FILL CONDITIONS IN SAN DIEGO, CALIFORNIA

Area Description, and Geologic and Soil Conditions

San Diego County is characterized by numerous mesas, canyons and valleys near the coast, and mountain ranges and desert terrain toward the east. Land values have become extremely high in this area of Southern California. High land values make it profitable to transform hilly sites into level sites by moving large volumes of soil, thus creating up to twice as many building sites. Because of this, area geotechnical engineers refer to the Southern California area as "the grading and earth moving capital of the United States." Fill thicknesses over 150 feet have been constructed in this region. Figure 6.2 shows a portion of a large grading operation in San Diego, where over 15 million cubic yards of fill is being placed.

High land values combined with a litigious environment have caused the integrity of a foundation and, subsequently, the structure to be more closely scrutinized. Homeowners who have invested large sums of money and are making very high mortgage payments have become extremely sensitive to cracks in the foundation and slabs. Often these cracks are concrete shrinkage cracks that do not necessarily affect the integrity of the foundation or the structure. But these conditions have forced engineers and designers to be ultra-conservative in their designs.

> High land values combined with a litigious environment have caused the integrity of a foundation and, subsequently, the structure to be more closely scrutinized. . . . These conditions have forced engineers and designers to be ultra-conservative in their designs.

Figure 6.2 Example of a large grading project in Southern California. (Photograph courtesy of Geocon Incorporated, San Diego, California.)

Common fill conditions in the San Diego area include deep compacted fills, uncontrolled fills, and cut/fill transitions. Each of these conditions is discussed within the following sections on the predominant foundation systems used in the San Diego area for fill sites.

Predominant Foundation Systems

Continuous wall and isolated footings in Southern California. The continuous wall and isolated footing system, built in conjunction with a concrete slab-on-grade, is undoubtedly the most common system in the San Diego area. Because of the popularity of this system, attempts are made to use it even when confronted with a site that will contain deep canyon fills.

The general design approach is to increase the depth of the continuous footings and the amount of longitudinal reinforcement as the total or differential thickness of the controlled fill increases. This creates a more rigid continuous footing that is able to undergo more stress due to differential movement of the soil. The amount of extra reinforcement required is based on empirical data and experience. Buildings constructed on isolated spread footings will still be susceptible to differential movement even if the amount of reinforcement is increased; buildings must be designed to withstand the predicted amount of movement.

The minimum continuous footing dimensions recommended for the area are 12 inches wide and 12 inches deep with two No. 4 bars, one near the bottom and one near the top of the footing. These dimensions are typically appropriate for sites that have relatively shallow depths of controlled fill (less than 20 feet thick). These minimum dimensions are also appropriate for sites that have less than 10 feet of differential controlled fill thickness.

For deeper fills, continuous footing recommendations are typically increased up to 24 inches deep and 12 inches wide, with four No. 4 bars, two near the bottom and two near the top of the footing. These dimensions are appropriate once total controlled fill thickness approaches 40 to 50 feet thick or greater. These recommendations also generally apply for sites that will have a differential controlled fill thickness of greater than 10 feet across the building pad.

For sites with deep controlled fills, floor slabs are generally reinforced with No. 3 bars spaced either 24 or 18 inches on center, in both directions. As the controlled fill becomes deeper (i.e., greater than 50 feet), the 18-inch spacing will typically be specified. Thicknesses of the slabs are typically 4 to 5 inches depending on their intended use (e.g., residential versus industrial warehouse).

Post-tensioned slab foundations in San Diego. As discussed in the beginning of this section on San Diego, there is great interest in controlling slab and foundation cracking due to shrinkage of the concrete, and/or differential movement due to settlement of the fill, in order to eliminate construction defect litigation. This has spurred an increase in the use of post-tensioned slabs for deep-fill conditions. Developers and builders are becoming more willing to incur the additional cost of the post-tensioning system in an attempt to reduce the potential for future lawsuits.

The design methodology of the post-tensioned slab was outlined in detail in Chapter 4. Briefly, steel reinforcing strands in the slab are tensioned after the concrete pour. Once the strands have been tensioned, steel plates at each end of the strands apply pressure to the slab edges as the strands attempt to return to their untensioned state. This, in effect, holds the slab together as it undergoes differential settlements and decreases the potential for shrinkage cracks to occur.

Chapter 4 also discussed some oversights in the design of the post-tensioned system in the San Diego area related to an unfamiliarity in the design process. This applies to the use of post-tensioned systems for deep and differential fills as well. The latest version of the *Uniform Building Code* (International Conference of Building Officials, 1988) has adopted the Post-Tensioning Institute's (PTI) design specifications as UBC Standard 29-4. However, many feel that this is an overly conservative design for the soil conditions in this region. The design manual does not specifically address the deep-fill application. Currently, to abide by the code, the geotechnical engineer must use the PTI guidelines to provide recommendations to the structural engineer and to follow the building code. The PTI method assumes minimum values of certain soil parameters. Occasionally, these parameters are higher than the actual site soil conditions. Therefore, to proceed with the design, the soil design values that must be used are worse than in reality. This leads to a more conservative design than is necessary. Structural and geotechnical professional groups in San Diego are currently preparing a design procedure with less stringent requirements to submit to the City Engineer. It is anticipated that an acceptable subsequent design will eventually be accepted for this region, which will include design procedures for deep-fill applications.

Drilled shaft foundations in the San Diego area. Occasionally, drilled shafts are used for commercial and industrial buildings on areas with uncontrolled fills in the San Diego area. They may also be used for large residential structures, such as an apartment building, or upscale residential homes that are to be constructed on a differential fill. The design methodology is to drill through the fill soils and to found the shaft on dense formational material. Grade beams will be used to support the walls of the structure, and the floor system will typically consist of a heavily reinforced concrete slab-on-grade.

The drilled shafts are typically 24- or 30-inch-diameter straight shafts. They are spaced relatively far apart (10 to 15 feet) to maximize the load carried by each shaft and thus minimize the number of shafts necessary. Any reinforcement in the shafts is intended for lateral support, because the shafts are generally not subjected to any large tensile forces.

Deep Engineered Fills in the San Diego Area

As described above, deep fills over 150 feet thick have been constructed in the San Diego area. Some of these fills have experienced significant settlements with time. It is believed that these settlements are related to poor compaction control and/or wetting-induced collapse (hydrocompaction). The solution to the former is to stress the importance of fill control and to maintain high standards during the fill process. Procedures such as removing all brush and vegetation from the fill soils, maintaining specified water contents, mixing the soil well enough to distribute the moisture in the soil uniformly, and testing the compaction of the soil on a regular and frequent basis that has been agreed upon by all parties involved before the fill process begins should be strictly adhered to.

Wetting-induced collapse of compacted soil occurs in some fills when the soil is wetted at high overburden stresses (Lawton et al., 1989). Although further research is being conducted on what types of soil are susceptible to wetting-induced collapse when compacted, preliminary results indicate that clayey soils appear to have this property. Under low overburden stress conditions these soils typically swell upon wetting (see Chapter 4). The relative compaction and water content used to compact the soil affects the potential for wetting-induced collapse. Higher relative compactions help to reduce the collapse of the compacted fill soil; however, they also increase the amount of expansion that may occur upon wetting even under high overburden stresses. Increased water content helps to reduce the potential for both collapse and expansion.

Lawton and others (1989) determined, for a given soil from a site in Southern California, that there is a critical relative compaction for a given overburden pressure where there was little to no volume change upon wetting. The critical relative compaction increases with increasing overburden pressure. It was also determined that the wetting-induced collapse was effectively reduced on samples that were compacted at or above the optimum water content for the specified compactive effort.

It is not practical to vary continually the relative compaction specification depending on the depth from proposed finish grade at which the engineered fill is being compacted. (However, many of the geotechnical firms in the San Diego area will specify a minimum relative compaction of 92% of the modified Proctor instead of a minimum relative compaction of 90% modified Proctor for fill soils that will end up deeper than 75 feet below proposed finished grade.) Therefore, it appears that the most practical method of reducing the potential for wetting-induced collapse of deep-fill soils is to specify that the soil must be compacted at, or above, the optimum water content.

LIGHTLY LOADED CONSTRUCTION ON LOOSE FILLS IN MIAMI, FLORIDA

Area Description, and Geologic and Soil Conditions

The predominant geologic formation in the Miami area is known as the Miami Limestone. This formation consists of limestone rock interstratified with dense sands. The soil conditions are typically suitable for continuous and/or isolated spread-footing foundations or thickened-edge monolithic slabs. The thickened edge acts as a continuous wall footing. It is thickened to help guard against erosion and the influx of moisture to the subgrade soil.

The area is relatively flat; therefore, large amounts of fill soil are typically not necessary to prepare a site for construction. However, along the coast of Miami, especially in the Miami Beach area, the majority of land consists of fills dumped under water. These fills were created by dredging the ocean bottom and dumping the material off the existing coastline. Consequently, these fill soils are very loose and typically overlie organic marine deposits.

Soil Improvement

A soil improvement technique known as vibroflotation is commonly used to densify the loose, sandy deposits and uncontrolled sandy fills. A vibroflot unit commonly consists of a vibrating jet that can be lowered into the soil. The jetting action causes the soil in the immediate vicinity temporarily to lose its strength, creating a condition in the soil that enables the weighted vibroflot to advance to a greater depth. The vibration of the vibroflot causes the granular material around it to densify. As it densifies, additional granular material is added from the ground surface in order to compensate for the volume decrease due to the densification of the soil.

Once the soil has been densified to the desired depth, either a monolithic slab foundation or a continuous-footing foundation is most commonly constructed. Compared to a deep foundation system, vibroflotation is a relatively inexpensive way to improve a loose sandy site for lightly loaded applications.

Pile Foundations in Miami

Thicknesses of hydraulically placed fill and marine organic deposits may approach 30 feet along the coastline in the Miami area. Two different types of pile foundations are occasionally used to minimize potential settlements in this region: precast concrete piles and auger cast piles. Pile foundations transfer the structural load through the loose fill soils to a more competent stratum below. Because of the vibrations caused by pile driving, the installation of the pile will typically tend to densify loose granular fills as well. Pile foundations in Miami are generally limited to commercial and upscale residential structures.

Precast piles. The precast concrete piles are typically 14 inches square. Concrete piles are preferred over steel and timber because of the higher cost of steel, the general lack of availability of timber piles in the southern Florida region, and the long-term durability of concrete relative to timber and steel. The piles are driven until a design driving resistance is achieved. This driving resistance is typically obtained at depths of 20 to 60 feet after the fill soil and marine organics have been penetrated and more competent formational material is encountered.

Auger cast piles. Auger cast piles are constructed using a continuous flight, hollow stem auger. In the Miami area the diameter of the flighting is typically 14 inches. The method of construction involves advancing the auger to the design depth for the pile. A cement grout is then pumped through the hollow auger into the soil. As the grout fills the void that is created by the auger and penetrates the surrounding soil, the auger is slowly withdrawn. The grout densifies the soil as it penetrates its loose structure under pressure. Figure 6.3 shows an auger cast rig in the process of drilling a pile for a proposed two-story commercial complex.

Grout pressures vary depending on the soil conditions. The *Augered Cast-In-Place Piles Manual* (Deep Foundation Institute, 1990) recommends maintaining a minimum pressure to produce a grout head of 5 feet. Readers are encouraged to refer to this document for additional information regarding auger cast piles. Reinforcing steel, as necessary, is either

Figure 6.3 Auger cast pile rig in Miami.

preloaded into the auger and centered with spacers or inserted into the pile immediately after the entire column has been grouted and the flighting has been withdrawn. Auger cast pile depths in the Miami area typically range from 20 to 60 feet for reasons similar to those for precast concrete piles.

Comparison of precast and auger cast piles. The auger cast construction method has several advantages over installing precast piles. These advantages include:

1. The cost of installing a typical auger cast pile is approximately equal to the cost of purchasing a precast concrete pile.

2. The auger cast method does not require the handling of long and awkward precast piles.
3. The auger cast piles are more readily adapted to changed site conditions. Depths can be altered by simply changing the depth of drilling.
4. Construction noise and vibrations are greatly reduced.

The auger cast system also has some disadvantages:

1. The grout volumes can become very large, especially if the soil is loosely packed or contains limestone solution voids (also common in Miami). Grout volumes can approach 2 to 3 times the volume of the augered hole, and it can sometimes take up to two concrete loads to fill them. This will also increase the cost of the pile.
2. The piles can become weakened if the wet grout is exposed to large amounts of organics while curing.
3. The ultimate load capacity of the pile is difficult to predict owing to the fact that it is difficult to characterize the soil/pile interaction because both the amount and the nature of the penetration of the grout into the soil are unknown.
4. The auger cast system is sensitive to construction procedures and therefore requires strict quality control and inspection.
5. Constant grout pressure must be maintained.

COMMON DESIGN AND CONSTRUCTION OVERSIGHTS

The quality of construction of an engineered fill is as important, and possibly even more important, than the choice of the foundation system for a lightly loaded structure. If the fill is not placed according to the grading specifications, the potential for differential settlement is increased. Attention must be given to the compaction specifications, choice of compaction equipment, and the nature of the soil to be compacted. E. J. Monahan has written an excellent practical text on soil compaction entitled *Construction Of and On Compacted Fills* (1986), which covers a vast amount of information in an easy-to-read format. This text is highly recommended for anyone who is professionally involved with earthwork. The following section includes observations from the Monahan text and some common oversights regarding constructing earth fills, as well as observations from other engineers and contractors around the country regarding fill and foundation construction and design oversights.

- **Modified or standard Proctor densities.** There are two commonly used laboratory tests that determine the maximum dry density and the corresponding optimum moisture content of a field sample of soil: the standard Proctor (ASTM D-698) and the modified Proctor (ASTM D-1557) tests. These laboratory tests consist of placing soil in lifts into a standard mold and compacting them by dropping a standard hammer a specific number of times. Once the mold is full it is weighed and the density of the material can be determined. A sample of the soil is dried to determine the moisture content by weight. The test is conducted at varying moisture contents in order to obtain a range of points along the compaction curve for the soil in question (see Chapter 2). The basic difference between the standard and the modified Proctor is the amount of energy that is put into the soil during the compaction process and the number of lifts or layers in which the soil is compacted. The modified Proctor energy is approximately 4.5 times greater than the energy of the standard Proctor test. Figure 2.6 graphically shows the difference between the two tests; additional information regarding compaction testing is presented in Chapter 2.

  The intended use of the standard Proctor was for the design of pavement subgrades. As time progressed the need arose to design fills that would support larger loads such as the impact loading of a jet landing or the load of a foundation footing. The modified test was established to better simulate the modern equipment being used to construct the airfields of World War II (Holtz and Kovacs, 1981), and building codes have since accepted the construction of foundations on fill using the modern equipment and subsequently higher soil densities. It is important *always* to specify whether the standard or modified Proctor test will be used for compaction control. Proper specification will minimize confusion regarding what type of compaction equipment and how many passes of the equipment will be necessary. Additionally, a fill designed for heavy loads may be underdesigned if a standard Proctor result is used to test a fill in the field. The vast majority of modern specifications reference the modified Proctor density. To avoid confusion it must still be made clear in the specification which Proctor test has been used.

- **Compaction equipment.** Consideration of the nature of the soil to be compacted is important in choosing the type of compaction equipment. Gravels and sands are compacted most efficiently by vibratory equipment. Clayey and silty soils are compacted most efficiently by using rolling or kneading equipment such as a sheepsfoot roller. Design densities may not be achievable with improper equipment.
- **Compacting lifts that are too thick.** Compaction is performed in layers (or lifts). The reason is that if all the fill soil were placed at once and compacted, the soil near the bottom of the fill would not be properly compacted. This is because the induced stress on the soil from the compaction equipment dissipates with depth. Figure 6.4 shows the distribution with depth of a load being applied to the soil surface. This is known as a *pressure bulb*. Its shape and magnitude is dependent on the size and weight of the compaction equipment (or the foundation footing dimensions and load). At a depth approximately twice the width (in cross section) of the compaction equipment the stress will have dissipated to approximately 10% of the original contact pressure at the ground surface as shown in Figure 6.4. Less than 10% of the contact pressure will have very little effect on the soil being compacted.

  Because the induced stress on the soil from the compaction equipment dissipates with depth, it is a good rule of thumb to not allow compaction lifts to exceed 12 inches in the uncompacted state. For some clayey or silty soil conditions this may be too thick, and the field technician or engineer should have the grading contractor adjust the lift thickness accordingly. The lift thickness specification should always be in terms of an uncompacted thickness. This is because the final compacted thickness is not known until it

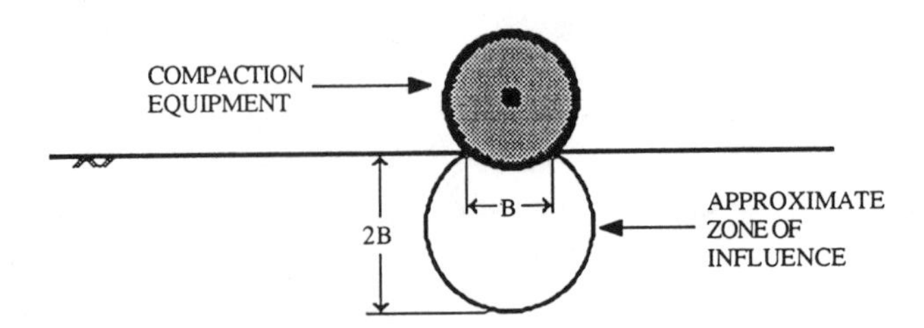

Figure 6.4 Approximate depth of influence of static compaction equipment on underlying soil.

has already been compacted. If the compacted lift is too thick, it is necessary to remove and recompact that lift.

It is important that the grading contractor, the site soil technician, and the design engineer understand the significance of the pressure bulb concept and lift thickness specifications. The surface of a compacted fill may be quite dense and appear to be properly compacted, yet 12 inches below the surface the soil may be in a very loose state and quite susceptible to future settlement.

- **Compaction and wet-soil conditions.** The presence of soft clayey soil, at natural water contents above the optimum moisture content for that soil, may tend to slow grading operations. The very moist, excavated soil should be allowed to dry or thoroughly mixed with drier soil in order to recompact it properly. In addition, it is difficult to achieve compaction specifications when compacting upon existing soils that have relatively high moisture contents; this is because the wet conditions do not provide a strong base upon which to compact the imported soils.
- **Routinely specifying 90% or 95% relative compaction.** Presently, the vast majority of compaction specifications are either 90% or 95% relative compaction of the maximum modified Proctor density. For the most part, this level of compaction is adequate for lightly loaded structures. Occasionally, conditions arise in which it may be advantageous to specify another compaction level. One example would be when the material is of an expansive nature. Compacting an expansive soil to a relatively high percentage of compaction on the dry side of the optimum moisture content increases the percentage, by volume, that the soil can swell. Thus, compacting expansive soils at a lower relative compaction, at or on the wet side of the optimum moisture content, may be advantageous as long as the strength of the soil does not remain too low. As an example, Figure 6.5 shows some recommended compaction specifications prepared by the Federal Housing Administration based on the plasticity index of the soil and the climatic rating for a specific region for slab-on-grade construction. Unfortunately, it is usually very difficult to convince local building departments to allow fill specifications below 90% relative compaction (modified Proctor) because 90% has become the industry standard of practice.
- **Neglecting the presence of oversize material in the soil to be compacted.** The oversize material in a soil sample is removed when

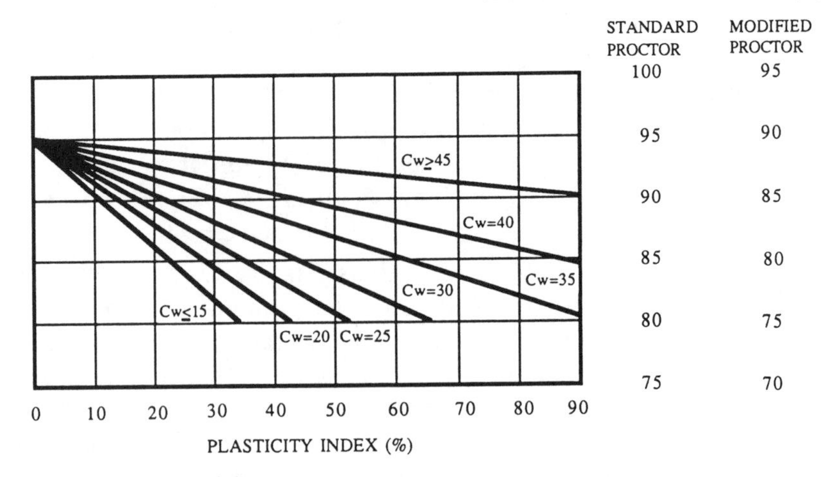

Figure 6.5 Optimum compaction requirements for slab site based on plasticity index and climatic rating. (From BRAB, 1968.)

a Proctor test is conducted in the laboratory. The size of the material that is removed depends on the test being done. For the sake of this discussion, oversize material will be considered to be that which is greater than 3/4 inch. The material is removed because of the limiting size of the mold in which the soil is compacted in the laboratory. Once this material has been removed, the Proctor test is conducted on the remaining soil, and the maximum dry density and optimum water content are both determined. However, in the field the oversize material is not removed. The result is that the maximum dry density and the optimum moisture content readings of the soil that contains the oversize material are different from that which was tested in the laboratory without the oversize material. This problem is typically resolved by using what is known as a *rock correction formula* to determine appropriate values for field control. The formula is dependent upon the laboratory Proctor results and the percentage of oversize material in the field. Day (1989) provides a discussion of various rock correction methods.

- **Neglecting the shape of the compaction curve.** As can be seen in Figure 6.6 different soil types tend to have different shaped curves. The shape of the curve affects the amount of field control that is necessary for moisture and energy levels. Silty soils are moisture sensitive, and clayey soils are energy sensitive (Monahan, 1986). It

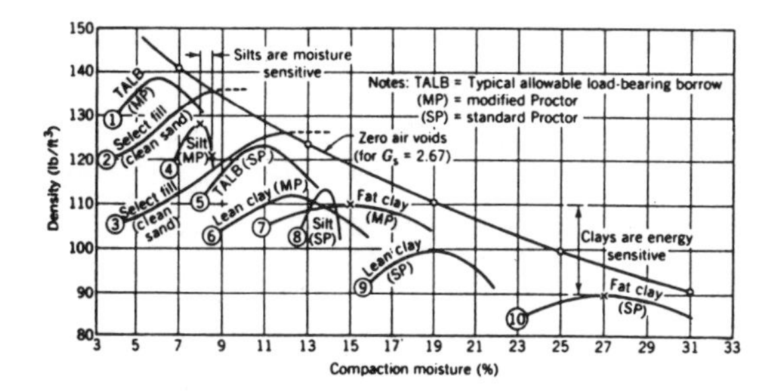

Figure 6.6 Variation of compaction curves for different soil types. (Reprinted by permission of John Wiley & Sons, Inc. Monahan. 1986. Copyright © 1986, John Wiley & Sons, Inc.)

can be inferred from the steepness of the silty soil curves that a slight deviation in the water content will lead to a large variation in dry density. Conversely, the flatness of the clayey curves suggests that changing the moisture level has a much less significant impact on the dry density. Changing the amount of energy that is put into the sample does have a large effect on clayey soils, potentially changing the maximum dry density by as much as 20 pounds per cubic foot, as shown in Figure 6.6. Therefore, it is important that the compaction energies are fairly uniform across a site in order to avoid future differential settlement of the fill. One example of how a large variation can occur is the routing of truck traffic on a site. If the construction vehicles are continually traversing the same course, the relative compaction beneath these temporary roadways can vary significantly from the rest of the site.

Figure 6.6 also illustrates that clean sands do not have an optimum moisture content. According to Monahan (1986), this phenomenon occurs because the sands are free draining; thus, when more water is added it drains out during the compaction procedure. Monahan (1986) recommends puddling these soils in order to compact them effectively. However, some governing agencies do not permit puddling as a method of compaction, and readers should check into the feasibility of using this method before proceeding.

- **Fill thicknesses.** If at all possible, fill thicknesses should be designed to be relatively uniform across a building pad. Otherwise, differential settlement should be expected and the foundation would need to be designed accordingly.

- **Cut/fill transitions.** Cut/fill transitions should not be designed to traverse beneath the proposed foundation of the building. The potential for differential settlement for such a condition is too great. If the situation does arise, it is generally recommended to undercut the lot by an additional 4 feet and replace it with compacted fill. This will provide a "cushion" for the building to settle upon. Otherwise it is possible that a "hinge" point will form where half the building is settling and the other half is not—owing to the presence of the formational soil.
- **Engineered fills will settle.** It is a common misconception that an engineered, compacted fill will not settle. This is not the case. The compaction greatly reduces the potential settlement, but it does not eliminate it. This is especially true for deeper fills. The weight of the fill itself may cause the soil to settle, as well as the future saturation of the fill soils from the influx of irrigation water. Engineers must incorporate a certain amount of settlement into the foundation design.
- **Downdrag of deep foundation members.** As an uncontrolled or loose fill settles, there is increased potential for downdrag forces to cause a driven or auger cast pile or a drilled shaft to fail. The settlement creates friction forces that act downward on any piles or shafts that pass through the fill. The downdrag forces may be large enough to cause the piles or shafts to settle excessively or fail because of overstressing.
- **Neglecting the history of the site.** Many settlement problems are related to old uncontrolled fills that were undetected during a geotechnical investigation. A simple check of a series of historical air photographs may reveal the placement of old fill. Talking to those who live or work in the area can also prove invaluable.
- **Scarifying the top 12 inches above a loose fill.** An engineer may specify the removal and recompaction of the top 12 inches of a loose fill in order to better support a foundation. Although this may be within the standards of the local building code, it may not be adequate for the building. The 12 inches of compacted soil will typically be strong enough to support the building in terms of a local bearing capacity failure of the footing; however, it is important to consider the pressure bulb concept. If a structure has a 12-inch-wide continuous wall footing, it will have a depth of influence of approximately 48 inches below the bottom of the footing. Similarly, a 24-inch-square

isolated spread footing will have a depth of influence of approximately 48 inches below the bottom of the footing. This influence may cause the loose fill beneath the thin compacted layer to settle.

- **Neglecting site variability when designing precast pile lengths through fills.** Often the subsurface conditions vary across a site that may have some loose fill near the surface. This may be a problem if precast concrete piles designed for a job are all one length. One area of the site could contain a thicker section of fill and require longer piles. It is extremely difficult to splice precast concrete piles together to create a longer pile. It is better to be conservative when specifying the necessary length of the piles and drive them deeper into the formational unit or cut them if refusal of the piles is encountered. Costs are often saved in the long run by driving test piles at the site (which will be incorporated into the foundation) before constructing all of the piles; this will afford a better idea of how long the remaining piles should be.
- **Creating a "quick" condition or inducing liquefaction during compaction.** Loose, saturated, or near-saturated granular soils may be susceptible to a sudden loss of strength if subjected to large vibrations. This phenomenon is sometimes referred to as a "quick" condition. The loss of strength may be large enough to disable compaction equipment. This is important to keep in mind when using vibrating compactors at sites where there is a relatively shallow water table.

SUMMARY

- The vast majority of engineered projects will involve fill soils.
- The four common fill conditions that affect lightly loaded construction are: (1) uncontrolled fills, (2) deep fills, (3) differential fills, and (4) cut/fill transitions.
- Footing depths and the amount of steel reinforcement are commonly increased for lightly loaded foundations constructed on deep fills in the San Diego area in an effort to strengthen the foundation in the event of subsequent differential settlement.
- Using post-tensioned slab foundations to reduce potential slab cracking due to settlement of fills is becoming popular in the San Diego area.

- Drilled shafts are occasionally used in San Diego to transfer foundation loads through differentially thick fills.
- Vibroflotation is used to densify uncontrolled granular fill soils in Miami.
- Both precast concrete piles and auger cast piles may be used in Miami at sites with loose, hydraulically placed fills along the coastline.
- It is extremely important to specify whether a standard or modified Proctor test has been used to determine the optimum moisture content and maximum dry density.
- Clayey soil is compacted most efficiently by using rolling or kneading compaction equipment. Granular soil is compacted most efficiently by using vibratory equipment.
- Compaction should be performed in lifts that are thin enough so that the compaction equipment can adequately compact the entire lift.
- If at all possible, cut/fill transitions should not be designed beneath a proposed building foundation. An alternative would be to undercut the lot by 4 feet and replace it with compacted fill.
- Engineered fills will experience settlement, especially if they are saturated in the future.
- Uncontrolled, loose, and engineered fills will create downdrag forces as they settle.
- A 12-inch compacted fill cap above a loose layer may not be sufficient to eliminate potential settlement and/or a bearing capacity failure.
- All engineered fills should be monitored by a qualified geotechnical engineering company.

RELATED REFERENCES

ADSC: The International Association of Foundation Drilling and DFI: Deep Foundation Institute, *Drilled Shaft Inspector's Manual*, 1st Edition, ADSC, Dallas, 1989.

Bowles, J. E., *Engineering Properties of Soils and Their Measurement*, 3rd Edition, McGraw-Hill Book Co., New York, 1986.

Brown, R. W., *Residential Foundations—Design, Behavior and Repair,* 2nd Edition, Van Nostrand Reinhold, New York, 1984.

BUILDING RESEARCH ADVISORY BOARD (BRAB) of the National Research Council, *Criteria for Selection and Design of Residential Slabs-on-Ground*, Report No. 33 to the Federal Housing Administration, National Academy of Sciences, Washington, DC, 1968.

COLEMAN, T. A., "Polystyrene Foam Is Competitive, Lightweight Fill," *Civil Engineering*, February 1974, pp. 68–69.

DAS, B. M., *Principles of Foundation Engineering*, PWS Publishers, Boston, 1984.

DAY, R. W., "Relative Compaction of Fill Having Oversize Particles," *Journal of Geotechnical Engineering*, ASCE, Vol. 115, No. 10, October 1989, pp. 1487–1491.

DEEP FOUNDATION INSTITUTE, *Augered Cast-In-Place Piles in Manual*, 1st Edition, Prepared by the Augered Cast-Place-Pile Committee, Deep Foundations Institute, Sparta, NJ, 1990.

DYWIDAG SYSTEMS INTERNATIONAL, *DYWIDAG Monostrand Posttensioning System*, 1990.

FLAATE, K., "The (Geo)Technique of Superlight Materials," *The Art and Science of Geotechnical Engineering At the Dawn of the Twenty-First Century*, Prentice-Hall, Englewood Cliffs, NJ, 1989.

FLEMING, W. G. K., ET AL., *Piling Engineering*, John Wiley & Sons, New York, 1985.

FRENCH, S. E., *Introduction to Soil Mechanics and Shallow Foundations Design*, Prentice-Hall, Englewood Cliffs, NJ, 1989.

FRYDENLUND, T. E., *Superlight Fill Materials*, Norwegian Road Research Laboratory, Publ. no. 60, 1986, pp. 11–14.

GREER, D. M., AND GARDNER, W. S., *Construction of Drilled Pier Foundations*, John Wiley & Sons, New York, 1986.

HAUSMANN, MANFRED R., *Engineering Principles of Ground Modification*, McGraw-Hill Book Co., New York, 1990.

HOLTZ, R. D., AND KOVACS, W. D., *An Introduction to Geotechnical Engineering*, Prentice-Hall, Englewood Cliffs, NJ, 1981.

INTERNATIONAL CONFERENCE OF BUILDING OFFICIALS, *Uniform Building Code*, ICBO, Whittier, CA, 1988.

LAMBE, T. W., AND WHITMAN, R. V., *Soil Mechanics*, John Wiley & Sons, New York, 1969.

LAWTON, E. C., FRAGASZI, F. J., AND HARDCASTLE, J. H., "Collapse of Compacted Clayey Sand," *Journal of the Geotechnical Engineering Division*, ASCE, Vol. 115, No. 9, September 1989, pp. 1252–1267.

LEONARDS, G. A., CUTTER, W. A., AND HOLTZ, R. D., "Dynamic Compaction of Granular Soils," *Journal of the Geotechnical Engineering Division*, ASCE, Vol. 106, No. 1, 1980, pp. 35–44.

MONAHAN, E. J., *Construction Of and On Compacted Fills,* John Wiley & Sons, New York, 1986.

NAHB RESEARCH FOUNDATION, INC., *Residential Concrete,* National Association of Home Builders, Washington, DC, 1983.

POST-TENSIONING INSTITUTE, *Design and Construction of Post-Tensioned Slabs-on-Ground,* 1st Edition, Post-Tensioning Institute, Phoenix, AZ, 1989. (a)

POST-TENSIONING INSTITUTE, *Field Procedures Manual for Unbonded Single Strand Tendons,* Post-Tensioning Institute, Phoenix, AZ, 1989. (b)

REESE, L. C., AND O'NEILL, M. W., *Criteria for the Design of Axially Loaded Drilled Shafts,* Center for Highway Research, Univ. of Texas, Austin, August 1971.

REESE, L. C., AND O'NEILL, M. W., *Drilled Shafts: Construction Procedures and Design Methods,* for the U.S. Dept. of Transportation, FHWA-HI-88-042, ADSC: The International Association of Foundation Drilling, ADSC-TL-4, Dallas, 1988.

SANDERS, G. A., *Light Building Construction,* Reston Publishing Co., Reston, VA, 1985.

SCHROEDER, W. L., *Soils in Construction,* John Wiley & Sons, New York, 1984.

SEED, H. B., "A Modern Approach to Soil Compaction," *Proceedings of the California Street & Highway Conference,* 11th Proceedings, 1959, pp. 77–93.

SEED, H. B., WOODWARD, R. J., AND LUNDGREN, R., "Prediction of Swelling Potential for Compacted Clays," *Journal of the Soil Mechanics and Foundations Division,* ASCE, Vol. 88, 1962.

SMITH, G. N., *Elements of Soil Mechanics,* 6th Edition, BSP Professional Books, Oxford, 1990.

TOMLINSON, M. J., *Pile Design and Construction Practice,* 3rd Edition, Palladian Publications, London, 1987.

U.S. DEPARTMENT OF THE INTERIOR, *Earth Manual,* A Water Resources Technical Publication, U.S. Dept. of the Interior, Bureau of Reclamation, Washington, DC, 1974.

ZEEVAERT, L., *Foundation Engineering For Difficult Subsoil Conditions,* 2nd Edition, Van Nostrand Reinhold, New York, 1983.

CHAPTER 7

Examples of Construction in Cold Weather Regions

INTRODUCTION

Cold weather can affect foundation construction when temperatures dip below freezing. Concrete and soil are affected by cold weather because they both contain water. For the purposes of this chapter, cold weather regions are defined as those that typically experience subfreezing temperatures for periods of several weeks to several months each year. This chapter does not discuss conditions of permafrost, where frozen conditions exist throughout the year.

As discussed in Chapter 2, some soils are susceptible to frost heave during prolonged periods of subfreezing temperatures. Frost heave may occur if there is a continual supply of moisture being drawn up by capillary action from the groundwater table. The forces created from the continued growth of ice lenses in the soil can be large enough to lift up lightly loaded structures. This can lead to differential movement of the foundation if the heave is not uniform, and it can damage the structure or crack the foundation. Additional damage may occur when the ice lenses begin to thaw. The soil typically thaws from the ground surface down. As the ice in the soil melts, the water cannot drain into the underlying ground that is still frozen. The soil may then settle owing to the increase in the water content and the loss of support of the now nonexistent ice

Table 7.1 Frost heave susceptibility of various soil types. (Adapted from Hansbo, 1975.)

Frost heave susceptibility	Soil type
None	Gravel and Sand
Moderate	"Fine" Clay (>40% clay)
Strong	Silt and "Coarse" Clay (clay content = 15–25%)

lenses. This settlement typically is nonuniform and leads to differential settlement of the foundation and damage to the structure.

Soils consisting primarily of silt-sized particles are the most susceptible to frost action. It is actually the size of the pore spaces (porosity) in the soil that controls the susceptibility to frost heave (Reed, Lovell, Altschaeffl, and Wood, 1979). Soil deposits that consist of up to 25% clay particles can exhibit a strong susceptibility to frost action if their porosity, and thus permeability, are relatively high (Hansbo, 1975). Although clay soils can exhibit high capillary rises, their permeability is relatively low and, therefore, they may not move the water fast enough to create significant ice lenses during a single freezing event. Conversely, silty soils can generate relatively high capillary rises in a relatively short time. This combination can lead to the formation of large ice lenses. Table 7.1 is a ranking of the susceptibility of different soils to frost heave and is adapted from Hansbo (1975).

Concrete is also affected by cold weather during construction. As temperatures decrease, the rate at which concrete gains strength also decreases. The cement in the concrete must react with the water in the concrete in order to build strength. If the water in the concrete freezes before the concrete reaches a compressive strength of 500 psi, it may be damaged beyond repair, even if it is subsequently thawed (American Concrete Institute [ACI], 1989). For this reason, concrete should never be placed directly on frozen ground. All snow, ice, and frost should be removed from the subgrade, forms, and reinforcing steel. If necessary, the subgrade should be covered and heated until all of the frost is melted. The ACI (1989) recommends heating the subgrade to 50°F, and concrete should be at a minimum temperature of 55°F when it is placed.

> As temperatures decrease, the rate at which concrete gains strength also decreases.

It is common to add air-entraining admixtures to the concrete mix to increase its durability against both frost action and the application of deicers on the concrete. Air-entraining admixtures cause small air bubbles to form inside the concrete. This procedure is known as *air-entrainment* (Wang and Salmon, 1985). For air-entrained concrete to be effective, it is recommended that 30 days of air drying take place after the concrete has cured for a minimum of 3 to 5 days before the use of deicers (ACI, 1989). The curing period consists of maintaining the moisture content in the concrete and keeping the temperature of the concrete between 40°F and 90°F (ACI, 1989). In general, deicers would not be used on a foundation, but this recommendation does apply to driveways, walkways, and garage slabs. Garage slabs are exposed to deicers as they drip off of a car.

If temperatures are below freezing, both the subgrade and the concrete need to be insulated to prevent freezing. Common insulating materials include fiberglass-filled blankets or straw. The insulation material should not be removed from the subgrade until the concrete is on-site and ready to be placed. Once the concrete has been poured, the insulation should be replaced on all exposed surfaces.

Another construction constraint during subfreezing temperatures is achieving proper compaction with frozen clayey and silty soils. Soils in this condition are difficult to work with and will lose strength once they thaw. Select fills consisting of sand and gravel mixtures are not as difficult to compact when they are frozen, and although importing soil adds to the cost of the project, they become a reliable substitute to frozen clayey and silty soils (Monahan, 1986).

Another construction constraint during subfreezing temperatures is achieving proper compaction with frozen clayey and silty soils.

The above constraints and phenomena affect the foundation design. The founding depth of the foundation is affected by the presence of subfreezing temperatures when considering the design of lightly loaded foundations. The *Uniform Building Code* (International Conference of Building Officials, 1988) specifies that it is necessary to found the foundation below the depth of frost penetration. Local building codes typically dictate the frost penetration depth. According to an article printed in *New England Builder* entitled ''Shallow Foundations Promoted'' (1988), many Scandinavian countries use shallow foundations above the frost line; these have been permanently insulated with foam to keep frost from penetrating beneath the slab. However, above-frost-line foundations have not yet gained acceptance by U.S. building codes.

There are different variations on how to accomplish the minimum founding depth, and there are some advantages of using one foundation system as opposed to another. The three regions that were a part of this study and that typically experience harsh winters include Denver, Colorado; Chicago, Illinois; and northeastern New Jersey. All three regions approach cold weather construction similarly. The remainder of this chapter uses examples from Chicago to illustrate foundation variations and construction considerations in cold weather conditions.

COLD WEATHER CONSIDERATIONS IN CHICAGO

Winters in Chicago can be extremely severe. The city's proximity to the Great Lakes and a large amount of yearly precipitation keep Chicago's groundwater table relatively shallow. As discussed in Chapter 5, varved silts and clay deposits are common in the Chicago area. These deposits consist of stratified layers of silts and clays that were deposited at the bottom of glacial lakes. The clay layers were deposited when the water beneath the frozen surface was calm in the winter, and the silt layers were deposited in the summer when the waters were more active, keeping the clay particles in suspension. The local building code requires a minimum foundation depth of 3.5 feet. For a site where the soil conditions dictate a deep foundation system, long-term problems associated with frost are not a primary consideration. This is because the foundation extends well beyond the frost penetration depth, and the amount of potential heave due to side friction is offset by the embedment of the deep foundation. For a site at which the soils have an adequate bearing capacity and a low potential for settlement, there are three general shallow foundation systems used to penetrate the frost depth: (1) continuous and/or isolated spread footings, (2) trench footings (a variation of the continuous footing), and (3) shallow drilled shafts. The following sections discuss the advantages and disadvantages associated with each of these approaches.

Continuous and/or Isolated Spread Footings

The advantage of the continuous and/or isolated spread footing method is that the majority of builders are familiar with it. However, several difficulties can be encountered during construction in subfreezing temperatures. If the ground is frozen it can be difficult to excavate with conventional equipment. Additionally, the on-site material may not be suitable for backfill if it is clayey or silty and becomes frozen. Once the excavation is completed it is necessary to keep the subgrade from freezing by using insulation material. Once a footing has been poured and formed, it is recommended that the forms be left on longer than usual because they act as additional insulation while the concrete is curing. Metal forms do not provide insulation and should be removed as soon as the concrete has set up. Once backfilled, the exposed portions of the footings should be kept insulated until the first-floor framing begins.

Trench Footing

The trench footing method consists of a trench, typically 8 to 12 inches wide, excavated down to the founding depth and then filled with concrete. If the soil conditions are conducive to this type of construction (i.e., high bearing capacity and will not collapse during the excavation) it can be a very inexpensive method of construction. Additionally, for frozen conditions, this type of foundation does not require any backfill, thus eliminating the need for clean, granular backfill. Excavations should be vertical and "sharp" near the top to avoid the formation of a lip during the concrete pour. It is possible that heaving of the soil could push up the foundation if a near horizontal lip is created.

Drilled Shafts

The drilled shaft method is one of the more simple systems to construct during the winter. Drilling through frozen ground is much simpler than excavating it with a backhoe. Similar to the trench foundation, there is no backfilling involved with the construction of the shafts. Drilled shafts are used most often during the winter as an alternative to isolated, interior footings. The diameter of the shaft is designed to create an equivalent-sized footing. For example, if a 2-foot-square footing were originally designed, creating a contact area of 4 square feet, an equivalent-sized shaft would have a diameter of approximately 27 inches.

(A 30-inch-diameter shaft would most likely be used due to common auger sizes.) The depth of the shafts would be identical to an equivalent isolated spread-footing design—deep enough to be founded below the frost penetration depth.

Basement Construction

Once constructed and heated, buildings with basements typically do not have problems with frost heave because their foundations extend below the frost depth. However, it is necessary to be aware of the potential for damage to occur during construction. Basement walls may be constructed during subfreezing temperatures, but the pouring of the floor slab may be postponed until the subgrade has completely thawed, to eliminate any potential cracking. If this is the case, it is a good idea to enclose and heat the basement to just above freezing. Otherwise it is possible for the soil on the exterior side of the walls to freeze and subsequently heave, damaging the wall, as the subfreezing temperatures penetrate the walls.

Subdrains are commonly installed to drain either a high groundwater table or to divert surface water from the subgrade soils in regions with expansive soils. The drains may outfall into a sump pit in the basement. The water is then pumped from the sump pit to a storm drain system. It is recommended that the sump pit be installed in such a manner that the owner of the structure can check it to ensure that the water has not frozen, which would not only damage the pump but would also back up the drain—potentially damaging the structure.

SUMMARY

- Subfreezing weather affects both the design and the construction of a foundation.
- Foundations constructed in regions where subfreezing temperature conditions commonly persist must be founded below the frost penetration depth in order to perform adequately.
- Soils that are predominantly silty are the most susceptible to frost heave.
- As frozen soil begins to thaw, the soil may settle as the water content of the soil increases and the support of the ice lenses dissipates.

- Concrete should not be allowed to freeze until it has reached a minimum compressive strength of 500 psi. Concrete should never be placed on frozen ground.
- Air-entrained concrete is more durable in subfreezing temperatures, but it should be air-dried at least 30 days before exposing it to deicers.
- Until the concrete has cured, and even as long as the building is under construction, foundations and the exposed subgrade should be insulated with straw, a fiberglass-filled insulation blanket, or other suitable insulation materials.
- Frozen clayey and silty soils are extremely difficult to compact. It is recommended that crushed stone or other clean granular materials be used as fill in subfreezing temperatures.
- Trench footings and drilled shafts are advantageous over conventional footings when constructing in subfreezing temperatures because they do not require soil backfill.
- Drilled shafts are occasionally used as a replacement for isolated, interior footings because of the additional effort necessary to excavate frozen soils.
- Basements should be enclosed and heated once the walls have been constructed so as to minimize the potential for the subgrade to heave.

RELATED REFERENCES

ADSC: The International Association of Foundation Drilling and DFI: Deep Foundation Institute, *Drilled Shaft Inspector's Manual*, 1st Edition, ADSC, Dallas, 1989.

AMERICAN CONCRETE INSTITUTE, *Guide to Residential Cast-in-Place Concrete Construction,* Report No. ACI 332R-84, American Concrete Institute, Detroit, 1989.

BOWLES, J. E., *Engineering Properties of Soils and Their Measurement*, 3rd Edition, McGraw-Hill Book Co., New York, 1986.

BUILDING RESEARCH ADVISORY BOARD OF THE NATIONAL RESEARCH COUNCIL, *Criteria for Selection and Design of Residential Slabs-on-Ground*, Report No. 33 to the Federal Housing Administration, National Academy of Sciences, Washington, DC, 1968.

FRENCH, S. E., *Introduction to Soil Mechanics and Shallow Foundations Design,* Prentice-Hall, Englewood Cliffs, NJ, 1989.

GREER, D. M., AND GARDNER, W. S., *Construction of Drilled Pier Foundations*, John Wiley & Sons, New York, 1986.

HANSBO, S., *Jordmateriallara,* Almqvist & Wiksell Forlag AB, Stockholm, 1975.

HOLTZ, R. D., AND KOVACS, W. D., *An Introduction to Geotechnical Engineering,* Prentice-Hall, Englewood Cliffs, NJ, 1981.

INTERNATIONAL CONFERENCE OF BUILDING OFFICIALS, *Uniform Building Code*, ICBO, Whittier, CA, 1988.

LAMBE, T. W., AND WHITMAN, R. V., *Soil Mechanics,* John Wiley & Sons, New York, 1969.

MONAHAN, E. J., *Construction Of and On Compacted Fills,* John Wiley & Sons, New York, 1986.

NAHB RESEARCH FOUNDATION, INC., *Residential Concrete,* National Association of Home Builders, Washington, DC, 1983.

NEW ENGLAND BUILDER, "Shallow Foundations Promoted," *New England Builder,* March 1988, pp. 8–9.

PORTLAND CEMENT ASSOCIATION, "Concrete Basements for Residential and Light Building Construction," *Concrete Information,* 1980.

PORTLAND CEMENT ASSOCIATION, *Joints in Walls Below Ground,* Concrete Report, Portland Cement Association, 1982.

REED, M. A., LOVELL, C. W., ALTSCHAEFFL, A. G., AND WOOD, L. E., "Frost Heaving Rate Predicted from Pore Size Distribution," *Canadian Geotechnical Journal,* Vol. 16, No. 3, 1979, pp. 453–472.

REESE, L. C., AND O'NEILL, M. W., *Criteria for the Design of Axially Loaded Drilled Shafts*, Center for Highway Research, Univ. of Texas, Austin, August 1971.

REESE, L. C., AND O'NEILL, M. W., *Drilled Shafts: Construction Procedures and Design Methods,* for the U.S. Dept. of Transportation, FHWA-HI-88-042, ADSC: The International Association of Foundation Drilling, ADSC-TL-4, Dallas, 1988.

SANDERS, G. A., *Light Building Construction,* Reston Publishing Co., Reston, VA, 1985.

SCHROEDER, W. L., *Soils in Construction,* John Wiley & Son, New York, 1984.

THOMPSON, H. P., "Supplementary Considerations for Slab-on-Grade Design," *Concrete International,* pp. 34–39, June 1989.

U.S. DEPARTMENT OF THE INTERIOR, *Earth Manual,* A Water Resources Technical Publication, U.S. Dept. of the Interior, Bureau of Reclamation, Washington, DC, 1974.

WANG, C. K., AND SALMON, C. G., *Reinforced Concrete Design,* 4th Edition, Harper & Row, Publishers, New York, 1985.

CHAPTER 8

Examples of Construction at Sites That Are Susceptible to Sinkhole Formation

INTRODUCTION

Approximately 15% of the land in the United States is underlain by rock formations that are susceptible to the formation of sinkholes (Beck and Sinclair, 1986). These regions are predominantly found in the Southeast, particularly in Florida, Alabama, Georgia, South Carolina, Kentucky, Tennessee, Missouri, as well as in Pennsylvania. Sinkholes are defined as "closed depressions in the land surface that are formed by solution of near-surface limestone and similar rocks and by subsidence or collapse of overlying surficial material into underlying solution cavities" (Beck and Sinclair, 1986). The phenomena typically occurs when acidic groundwater is in contact with limestone rock for a prolonged period of time. The acidic water dissolves the limestone as it moves through the pores or over the top of the rock, creating increasingly larger voids and cavities.

Sinkholes are defined as "closed depressions in the land surface that are formed by solution of near-surface limestone and similar rocks and by subsidence or collapse of overlying surficial material into underlying solution cavities."

The process that creates a sinkhole is slow and continuous. However, the effects of the sinkhole at the ground surface may occur either catastrophically or slowly over time. These two different sinkhole occurrences are referred to as *collapse sinkholes* and *subsidence sinkholes*, respectively. Collapse sinkholes occur when the solution of the limestone along a crack or a joint creates a predominantly vertical cavern or "throat" beneath the ground surface. At first the soil at the surface may be strong enough to bridge the cavern. With time, however, the cavern will continue to widen, and the bridging-surface soils will eventually collapse. This catastrophic or collapse-type of sinkhole is the kind that most people have heard about in media reports. Figure 8.1 shows a photograph of a large collapse sinkhole that occurred in 1981 in Winter Park, Florida; Figure 8.2 shows diagrammatically how this type of sinkhole typically forms.

Figure 8.1 Sinkhole in Winter Park, Florida, May 1981. (Reprinted by permission of ©The *Orlando Sentinel*, photographed by George Remaine, May 15, 1981.)

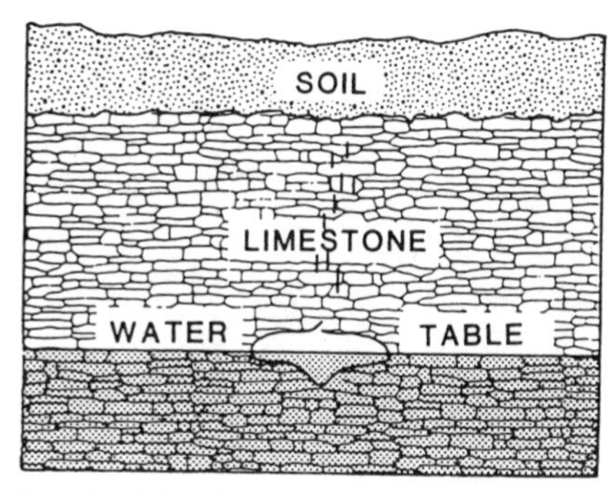

(a - above) Solution cavity develops along joint or other plane of weakness at the water table.

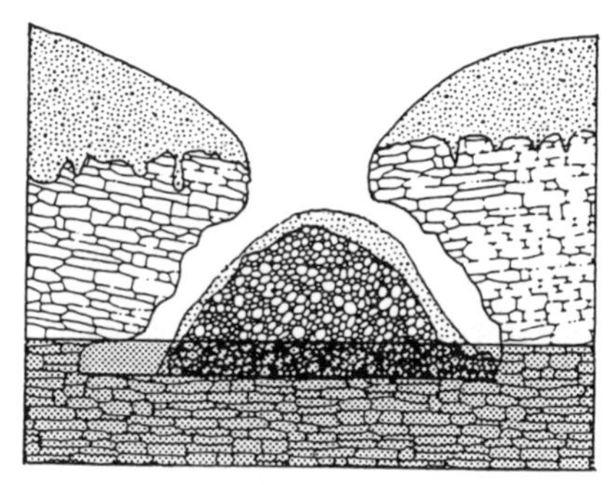

(c - above) Roof collapse reaches land surface. Undercutting continues.

(b - below) Roof collapses, most likley at joint intersection. Undercutting of cave walls by diverted groundwater.

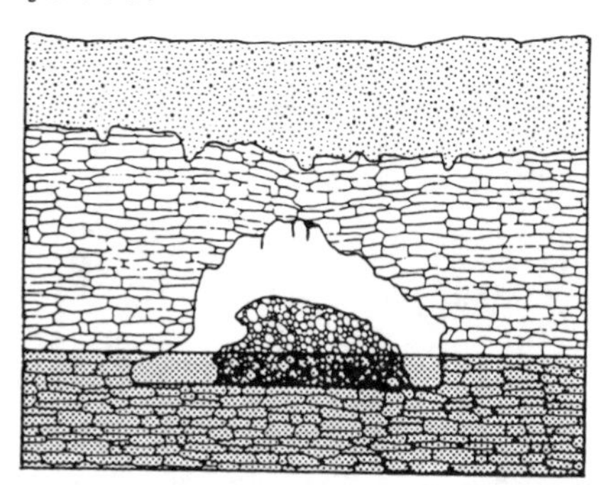

(d - below) Soil washes into depression and obscures its origin. Breakdown and cave roof cemented by recrystallized limestone. Groundwater flow established in new conduit.

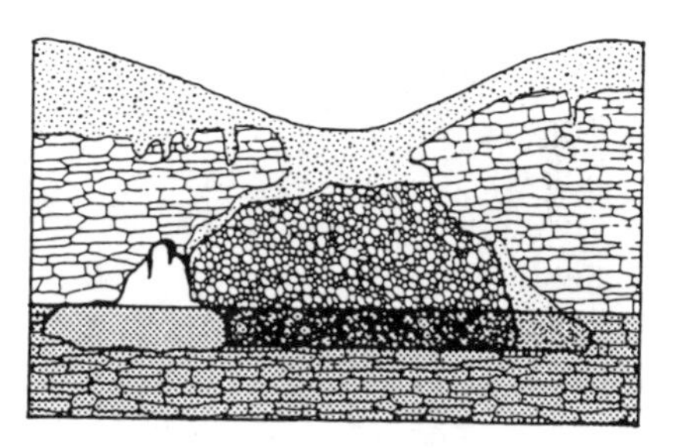

Figure 8.2 Stages in development of a collapsed sinkhole. (Reprinted by permission of the Florida Sinkhole Institute. From Beck and Sinclair, 1986.)

Subsidence sinkholes occur when the soil above the limestone formation is relatively granular. As the limestone erodes, the soil above it fills the voids. This is referred to as *ravelling* of the soil. As the soil continues to ravel into the limestone voids, the ground surface will begin to subside, forming an increasingly growing sinkhole.

Another surface feature associated with the solution of near-surface limestone deposits are *solution voids.* These are smaller features that may occur from surface water percolating into the joints of an exposed limestone layer. With time, they may be filled up with loose sandy deposits, which can hide their presence. These features are difficult to identify during the field investigation because of their small size. This condition,

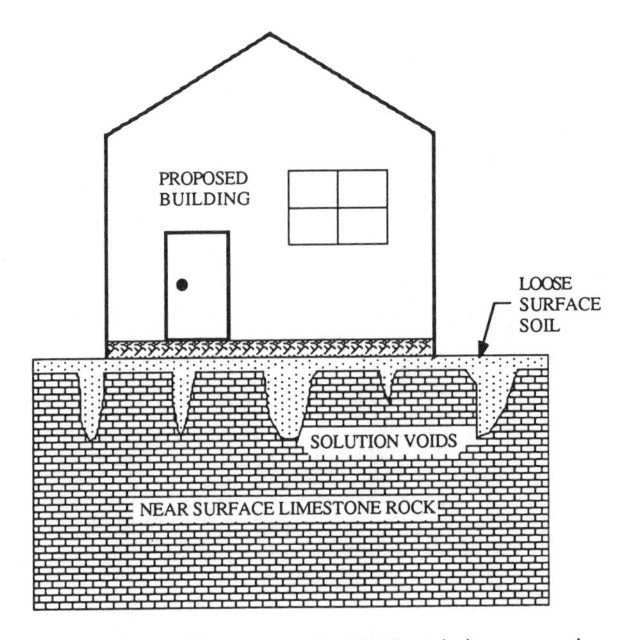

Figure 8.3 Solution voids filled with loose sand.

if undetected, may lead to foundation problems in the future. Figure 8.3 shows the characteristics of solution voids.

Sinkholes vary in size depending on the thickness of the overlying stratum. If the overlying stratum is relatively thick, it can span a larger cavern. Therefore, the cavern must grow larger in order to cause a collapse—thus creating a larger sinkhole. Sinkholes can reach sizes that can "swallow" entire structures. The sinkhole shown in Figure 8.1 from Winter Park, Florida, was approximately 300 feet in diameter at the ground surface and over 100 feet deep (Beck, 1988). A house and three cars were lost in this sinkhole.

The presence of urban development can often accelerate the formation of sinkholes or limestone solution voids. Changes in the environment that can trigger an acceleration include: (1) road construction, (2) landscape and vegetation changes, (3) building construction, and (4) groundwater pumping. Road construction can cause changes in drainage patterns. The new drainage patterns may divert water into a sinkhole-susceptible area and accelerate the solution of the limestone (White, Aron, and White, 1984). Changing the landscaping can affect the rate of limestone solution if the new landscape requires an increase in irrigation. The additional weight of a building may also cause the soil over an existing limestone throat to collapse. Urban and/or irrigation pumping of the groundwater can cause a sinkhole to collapse from increased groundwater flow rates through the soil and rock. The pumping can also cause seasonal fluctuations in the groundwater level (Newton, 1984). The lowering of the groundwater decreases the amount of buoyant support provided to the soil, effectively causing the soil to become heavier and leading to a higher probability that the soil will collapse into an underlying void or cavern.

SINKHOLE DETECTION

Detecting potential sinkholes that have not yet affected the ground surface can be relatively difficult. The traditional geotechnical investigation consisting of drilling boreholes may not detect potential sinkholes because a borehole samples only a small area. Benson and La Fountain (1984) suggest that utilizing borings is only 10% to 20% accurate in terms of locating sinkholes. Several other methods are commonly used to aid the search for potential sinkholes. These include: (1) looking at historical aerial or satellite photographs, (2) backhoe trenches, and (3) various geophysical techniques. These methods are examined in the following sections.

Aerial and Satellite Photography

Aerial photographs can aid in locating potential sinkholes. Historical air photos may show large-scale areas of subsidence or they may show "pock" marks or small lakes that can indicate smaller-scale localized sinkholes. These features are recognizable from the ground but may be much more readily identified from the air. Historical photographs may also reveal subsided areas in the past that are now filled with import material. Satellite photographs can reveal similar surface features associated with sinkhole development on a larger scale. Another source of historical information is the U.S. Geologic Survey, which has compiled regional maps showing areas of sinkhole potential and occurrence.

Backhoe Trenches

Backhoe trenches are useful simply because a relatively large area (compared to a borehole) can be explored relatively quickly. Backhoe trenches can expose near-surface solution voids or developing sinkhole throats. They cannot completely replace information that can be obtained from a borehole (for example, in situ densities), and they are also limited in the depth to which they can readily explore. Because trenches cannot completely replace borings it is sometimes difficult to convince the average developer/builder that utilizing both borings and trenches as part of the site investigation is worth the extra cost.

Geophysical Surveys

Various geophysical techniques can be used to locate cavities and sinkhole throats. It is beyond the scope of this text to discuss each method. The general idea behind geophysical techniques is to probe the subsurface without disturbing the ground surface. This is accomplished by generating a wave, which, when propagated through the soil, will reveal the presence of a cavity. Examples of such surveys include seismic waves, electrical resistivity, and ground-penetrating radar. A description of these techniques and other geophysical techniques can be found in Benson and La Fountain (1984).

SOIL TREATMENTS AND FOUNDATION SYSTEMS

According to Sowers (1984), there are five courses of action that may be selected to mitigate potential foundation problems attributable to the presence of a sinkhole(s) on the site. These are:

1. Optimize the location on the site.
2. Correct or mitigate defects [in the subsurface] that are present.
3. Use modified shallow foundations to overcome the defects [in the subsurface].
4. Use deep foundations to overcome the defects [in the subsurface].
5. Minimize future activation of [the] defects [in the subsurface].

Other than the first and last, these mitigation techniques are typically only used for commercial and industrial buildings and upscale residential structures. The reason for this is that the mitigation techniques are generally too expensive for tract-home budgets. What follows is a discussion of the different techniques listed above in terms of the type of lightly loaded construction.

Residential Buildings

When it has been determined that the potential for future sinkhole formation exists at a proposed residential development, it is somewhat rare for anything to be done to mitigate the potential during construction. The reason for this generally revolves around cost and risk assessment. Although the potential for sinkhole development may exist, it is generally unknown whether it will take one year or one hundred years for the sinkhole to become active at the ground surface. Therefore, unless there is a known active sinkhole, it may not be worthwhile for the builder to spend $20,000—for a structure that will have a value of $100,000—either to alter the subsurface conditions or to construct a foundation that will provide support in the event of a sinkhole forming.

Although the potential for sinkhole development may exist, it is generally unknown whether it will take one year or one hundred years for the sinkhole to become active at the ground surface.

If there is an active sinkhole discovered at the ground surface, the decision not to build on this site may be made for a residential development. If the sinkhole is relatively small (2 to 3 feet in diameter) compared to the size of the lot, the location of the lot and the house may be optimized in such a way that the likelihood that the sinkhole will have any effect is reduced. Typically, the lot will be designed such that the cavity is located in the backyard if possible. The homeowner must then either live with, or continually attempt to fill in the ever-present hole on the lot. If the sinkhole covers a large portion of the lot it becomes economically unfeasible to build a home on it.

Another economic factor affecting residential construction in areas of sinkhole potential is insurance. Some states now subsidize insurance against damage done by sinkholes. As an example, Florida has had a state-subsidized insurance program for residential structures damaged by sinkholes since 1981. Insurance programs such as these lessen the risk to the builder and, from the builder's perspective, make any type of soil treatment not worth the additional expense.

Precautions can be taken to attempt to minimize the rate at which any known sinkholes will grow or the rate at which surface soil will ravel into the cavity. Site drainage is the predominant method of control. It is essential to maintain positive drainage away from the sinkhole area. Groundwater pumping in the area should be minimized. The lowering of the groundwater table can cause limestone solution rates to increase as the flow regime of the groundwater is changed.

Commercial and Industrial Buildings

It is much more common to see soil treatments or special foundation systems for commercial structures than for residential structures. This is because the higher value of the commercial structure makes the cost of soil treatment or a special foundation system economically feasible. Some of the more commonly used alternatives for commercial and industrial buildings are listed below:

1. To correct or mitigate the cavities in the subsurface that are present, which includes procedures such as (a) grouting up the throat or cavity, (b) cap grouting, and (c) dynamic compaction.
2. To use modified shallow foundations to overcome the cavities in the subsurface; this includes (a) creating a footing that spans the cavity and (b) constructing a mat foundation that is rigid enough to minimize deflections that may occur due to sinkhole formation beneath it.
3. To use deep foundations, such as driven piles or drilled shafts to bypass the cavities in the subsurface.

Throat or cavity grouting. Throat grouting is a process of sealing up the throat and/or voids with a cement grout or concrete in order to prevent further ravelling of surface or near-surface soils into the limestone throat and to create a bearing surface for the foundation. Once grouted, the throat or cavity may be capped with an impervious soil stratum to prevent the influx of water. Cutting off any water reduces the potential for the throat to open up again as the limestone continues to dissolve. Throat grouting can become expensive if the cavities are large. An estimate of the size of the throats should be made and the costs of filling them up should be compared with alternative foundation solutions such as drilled shafts or driven piles.

Cap grouting. Cap grouting may be used to attempt to cover or bridge a series of sinkholes or cavities. The grouting is carried out by drilling down to the limestone stratum beneath the location of one of the proposed building columns and pumping a low-slump cement grout into the hole. If the grout take is large it is assumed that a cavern has been located. The grout is pumped under low pressure in an attempt to keep the grout localized in the stratum above the cavern. The idea behind cap grouting is to bridge the caverns, not fill them up. The grout bridge (or cap) stops any further ravelling of the soil above the caverns. Sometimes the grout is allowed to harden overnight to create a hard surface upon which the grouting operation is continued the following day.

If a cavern is located, the locations beneath the adjacent building columns will be drilled and grouted in order to attempt to extend the bridge. If the grout take is small it is assumed that a cavern has not been located and the adjacent column locations will then be skipped. This process is continued until all of the column locations within the building pad have either been grouted or it has been determined that there is not a cavity at that location.

Cap grouting is a method that attempts to prevent further sinkhole formation and the ravelling of surface soils. It does not add significant strength to the soil. Therefore, the support characteristics and compressibility of the surface soil must still be considered.

The effectiveness of cap grouting is difficult to evaluate for two reasons. The first reason is it is difficult to assess if the grouting has indeed covered the entire cavity or series of voids in the limestone. The second reason is the majority of cap-grouted sites have been sites where the voids in the limestone were already present but not necessarily currently active. Therefore, it is not known whether cap grouting has been effective at these sites or whether the solution process is simply inactive (Sowers, 1984).

Dynamic compaction. Dynamic compaction consists of dropping a very heavy weight on the surface of the soil. When implemented successfully, the impact of the weight causes the limestone cavities and throats to collapse. It also densifies any loose soils that have ravelled into the cavities. The weights can range up to 220 tons; however, they are more commonly in the 10- to 45-ton range (Holtz and Kovacs, 1981). Drop heights commonly range from 30 to 125 feet. According to Leonards, Cutter, and Holtz (1980), the depth of influence of dynamic compaction is related to the height of the drop and the magnitude of the weight. In some cases, depths up to 60 feet can be improved economically (Guyot, 1984).

Once the dynamic compaction is completed, it is necessary to fill in the "craters" that are created from dropping the weight. These craters should be carefully backfilled as discussed in Chapter 6. Blasting is a similar method to dynamic compaction in that the objective is to collapse any cavities.

Dynamic compaction does not necessarily impede surface water seepage nor eliminate the flow of groundwater. Therefore, it is possible that the flow of water could cause new cavities to form by dissolving the collapsed limestone rock.

Constructing a modified shallow foundation system. This method of construction consists of designing a footing that is rigid and strong enough to span across the projected area of the suspect sinkhole. The foundation must be strong enough so as not to deflect excessively in the event that the void becomes larger. A building of this type usually requires a monitoring program to ensure that deflections of the slab or foundation are not approaching excessive levels (Sowers, 1984). Floor slabs can be similarly reinforced to bridge suspected sinkhole areas. The structure should also be made sufficiently rigid enough so it will not be distressed in the event that the foundation does undergo some minor deflections.

Deep foundation systems

Driven Piles. The most common deep foundation solution is to drive piles through the cavernous layer into more competent material below. Occasionally auger cast piles will be used if the voids in the limestone are relatively small (e.g., limestone solution voids versus large sinkhole formations). Auger cast piles are constructed by pumping cement grout through a continuous-flight hollow-stem auger that has been drilled to the design depth. Auger cast pile systems are described in more detail in Chapter 6.

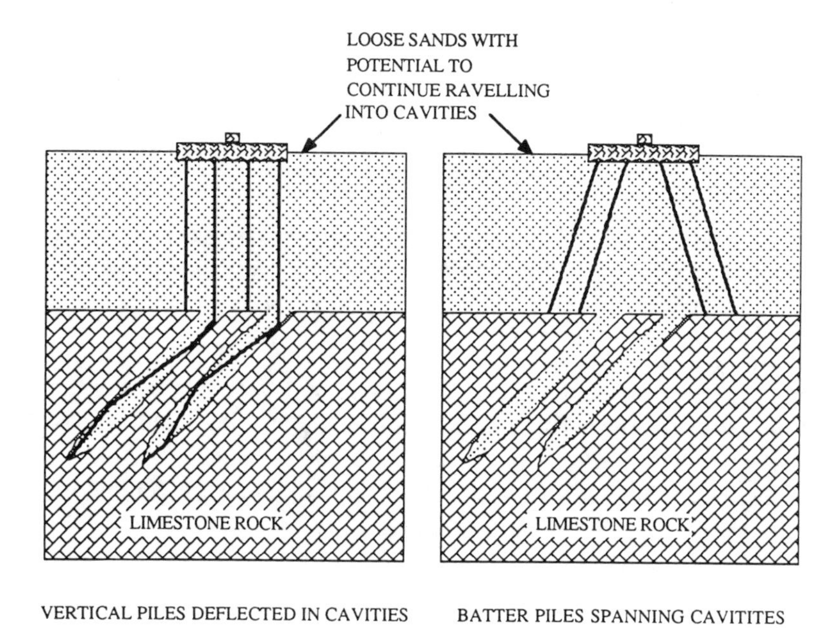

Figure 8.4 Driven pile configurations in limestone terrane. (Adapted from Sowers, 1984.)

A disadvantage of driving piles is that it is possible for a pile to deflect on a sloping surface of the limestone rock and follow a cavity during the driving, as shown in Figure 8.4. This can lead to very long piles that may not provide adequate bearing (Sowers, 1984). One alternative is to span a known void with batter piles (piles driven at an angle) as shown in Figure 8.4. This is effective only if the extent of the void is well defined; for example, by using ground-penetrating radar.

Drilled Shafts. Drilled shafts are used as an alternative to driven piles and are employed to penetrate the weak soils and the limestone zone containing solution cavities. Casing is typically required because of the presence of loose soils that commonly fill the cavities and the presence of groundwater. The drilled shaft has the advantage over the pile system in that the penetration depth of the foundation is known. In contrast, a pile may be deflected or damaged during driving.

The disadvantage of using a deep foundation system for an area with sinkhole potential is that the foundation system does not arrest the growth of the sinkhole. The building will be readily supported by the shafts or piles but nothing has been done to stop the growth of the sinkhole. Therefore, it is conceivable that soil may continue to ravel into the throat of the sinkhole and eventually the building may be situated on stilts.

ADVANTAGES AND DISADVANTAGES OF DIFFERENT SOIL TREATMENTS AND FOUNDATION SYSTEMS FOR MITIGATING SINKHOLE-SUSCEPTIBLE SITES

The table below summarizes the advantages and disadvantages of the different approaches to mitigating sites susceptible to sinkhole formation or limestone solution voids that were presented in this chapter. Additional approaches are available; however, those listed below are commonly implemented for lightly loaded structures.

Method	Advantages	Disadvantages
Optimization	Relatively inexpensive. Appropriate for residential construction.	Does not arrest sinkhole formation. May not be feasible if lot size is small. Continual maintenance of hole may be required.
Minimize Potential for Future Sinkhole Activation	Relatively inexpensive. Appropriate for residential construction.	Appropriate only if sinkhole is known to be currently inactive.
Correct Voids in the Limestone Substratum	Can arrest future additional sinkhole formation if throat is sealed. Shallow foundations may be utilized.	Effectiveness of grouting is unknown. Dynamic compaction does not arrest sinkhole formation.
Rigid Shallow Foundations	Cost is not prohibitive.	Requires a more rigid structure as well as the foundation. Requires periodic monitoring of the building to ensure that deflections are not approaching failure levels due to enlarging sinkhole.
Deep Foundations	Supports structure even if sinkhole enlarges.	Higher up-front construction cost. Condition of tip of driven pile is unknown.

SUMMARY

- Approximately 15% of the land in the United States is susceptible to sinkhole formation.
- Sinkholes are formed by the solution of carbonate rocks from the chemical reaction with acidic groundwater.
- There are two general types of sinkholes: collapse and subsidence. Collapse sinkholes occur catastrophically when a cavern created by the solution of the rock becomes so wide that the overlying soil cannot bridge it any longer. Subsidence sinkholes occur with time as the overlying granular soils ravel into a cavern or void.
- Urban development may accelerate sinkhole formation by altering drainage paths.
- Because of the limited site coverage provided from soil borings, sinkholes can be relatively difficult to detect if soil borings are the only tool used during the subsurface investigation. Geophysical techniques, historical aerial photos, and backhoe trenches are typically used to enhance soil-boring information if sinkhole activity is suspected.
- For economic reasons, soil treatment or special foundations are not typically used to protect residential structures from potential sinkhole damage.
- Commercial sites are typically treated in order to reduce the potential for a sinkhole to damage the structure. Once the soil/sinkhole treatment is complete, shallow foundations such as continuous spread footings or a thickened-edge monolithic slab become appropriate.
- Deep foundations may be utilized to support the structure but this method does not remove the potential for a sinkhole to undermine the soil beneath the structure.

RELATED REFERENCES

AMERICAN SOCIETY OF CIVIL ENGINEERS, *Geotechnical Aspects of Karst Terrains*, Proceedings of a symposium sponsored by the Committee on Engineering Geology, ASCE, New York, 1988.

BECK, B. F., AND SINCLAIR, W. C., *Sinkholes In Florida: An Introduction,* Report 85-86-4, The Florida Sinkhole Research Institute and The U.S. Geological Survey, 1986.

BECK, B.F., "Environmental and Engineering Effects of Sinkholes–the Processes Behind the Problems," *Environ. Geol. Water Sci.* Vol. 12, No. 2, 1988, pp. 71–78.

BECK, B. F., "Sinkhole Terminology," in *Sinkholes: Their Geology, Engineering and Environmental Impact,* Proceedings of the First Multidisciplinary Conference on Sinkholes, Orlando, FL, 1984.

BENSON, R. C., AND LA FOUNTAIN, L. J., "Evaluation of Subsidence or Collapse Potential Due to Subsurface Cavities," in *Sinkholes: Their Geology, Engineering and Environmental Impact,* Proceedings of the First Multidisciplinary Conference on Sinkholes, Orlando, FL, 1984.

BENSON, R. C., AND YUHR, L. B., "Assessment and Long Term Monitoring of Localized Subsidence Using Ground Penetrating Radar," in *Karst Hydrogeology: Engineering and Environmental Applications,* Proceedings of a conference sponsored by the Florida Sinkhole Research Institute, College of Engineering, University of Central Florida, Orlando, FL, 1987.

GARLANGER, J. E., "Foundation Engineering in Deeply Buried Karst," *The Art and Science of Geotechnical Engineering At the Dawn of the Twenty-First Century,* Prentice-Hall, Englewood Cliffs, NJ, 1989.

GUYOT, C. A., "Collapse and Compaction of Sinkholes by Dynamic Compaction," in *Sinkholes: Their Geology, Engineering and Environmental Impact,* Proceedings of the First Multidisciplinary Conference on Sinkholes, Orlando, FL, 1984, pp. 419–423.

HAUSMANN, MANFRED R., *Engineering Principles of Ground Modification,* McGraw-Hill Book Co., New York, NY, 1990.

HOLTZ, R. D., AND KOVACS, W. D., *An Introduction to Geotechnical Engineering,* Prentice-Hall, Englewood Cliffs, NJ, 1981.

LEONARDS, G. A., CUTTER, W. A., AND HOLTZ, R. D., "Dynamic Compaction of Granular Soils," *Journal of the Geotechnical Engineering Division,* ASCE, Vol. 106, No. 1, 1980, pp. 35–44.

NEWTON, J. G., "Review of Induced Sinkhole Development," in *Sinkholes: Their Geology, Engineering and Environmental Impact,* Proceedings of the First Multidisciplinary Conference on Sinkholes, Orlando, FL, 1984.

SOWERS, G. F., "Correction and Protection in Limestone Terrane," in *Sinkholes: Their Geology, Engineering and Environmental Impact,* Proceedings of the First Multidisciplinary Conference on Sinkholes, Orlando, FL, 1984, pp. 373–378.

STANGLAND, H. G., AND KUO, S. S., "Use of Ground Penetrating Radar Techniques to Aid in Site Selection for Land Application Sites," in *Karst Hydro-*

geology: Engineering and Environmental Applications, Proceedings of a conference sponsored by the Florida Sinkhole Research Institute, College of Engineering, University of Central Florida, Orlando, FL, 1987.

WHITE, E. L., ARON, G. AND WHITE, W. B., "The Influence of Urbanization on Sinkhole Development in Central Pennsylvania," in *Sinkholes: Their Geology, Engineering and Environmental Impact,* Proceedings of the First Multidisciplinary Conference on Sinkholes, Orlando, FL, 1984.

YOKEL, F. Y., "Housing Construction in Areas of Mine Subsidence," *Journal of the Geotechnical Engineering Division,* ASCE, Vol. 108, No. GT9, September 1982, pp. 1133–1149.

CHAPTER 9

Examples of Construction on or Near Slopes

INTRODUCTION

As discussed in Chapter 2, slopes and hills have an inherent stability problem. The driving force affecting their stability is gravity. Gravity is constantly "pulling" on the slope in an attempt to force it into a more stable configuration. The stability of a slope is affected by the soil type and strength, the presence of the groundwater table, the presence of a weak subsurface layer, and the geometry of the slope. Slopes can fail due to a variety of reasons. Examples include: (1) their own weight, (2) an influx of water, which may reduce the shear resistance of the soil, (3) dynamic loading from an earthquake, (4) erosive action, and (5) undercutting of the slope owing to construction. Additionally the constant pull of gravity can cause a slope to move slowly downward and outward with time, affecting structures constructed at the top of it. This is commonly referred to as *creep* and is more common on slopes with clayey soils because of their plastic nature.

The construction of a structure or the addition of fill on or near the top of a slope will increase the probability of these types of failures. The increase in probability is due to the additional weight that the slope must now support. Additionally, irrigation of the landscaping around the new structure will decrease the stability of the slope

> **The stability of a slope is affected by the soil type and strength, the presence of the groundwater table, the presence of a weak subsurface layer, and the geometry of the slope.**

because the addition of water may weaken the shear resistance of the soil.

There are two general types of slope failures: rotational and sliding. A rotational failure typically occurs along a circular surface, as shown in Figure 9.1. The location of the potential failure surface depends on the geometry of the slope, the strength and unit weight of the soil, the location of any surface loads, and the presence of any groundwater. Any change in these parameters will cause the location of the potential failure surface to change.

A sliding failure commonly occurs when there is a weak stratum or seam of soil within the slope, as shown in Figure 9.2. If the shear resistance of the weak stratum is less than the driving force due to gravity, it is likely that the weak stratum will fail, bringing the overlying soil with it. The weak material commonly consists of a thin seam of clay or weathered shale.

A third type of slope movement consists of a relatively slow phenomenon (commonly referred to as "creep") that causes an entire slope

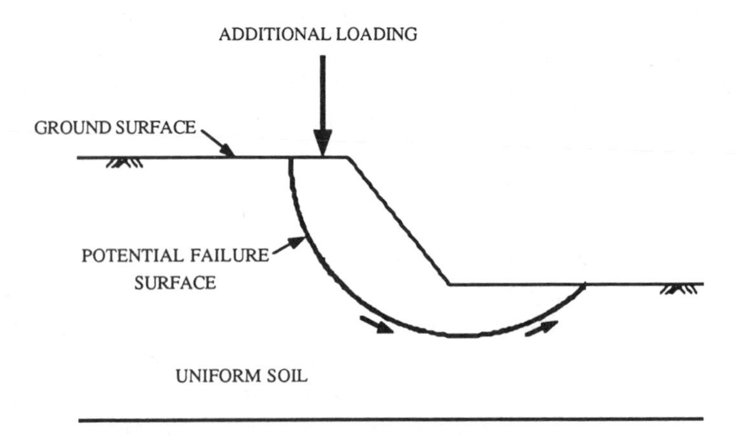

Figure 9.1 Circular potential failure surface.

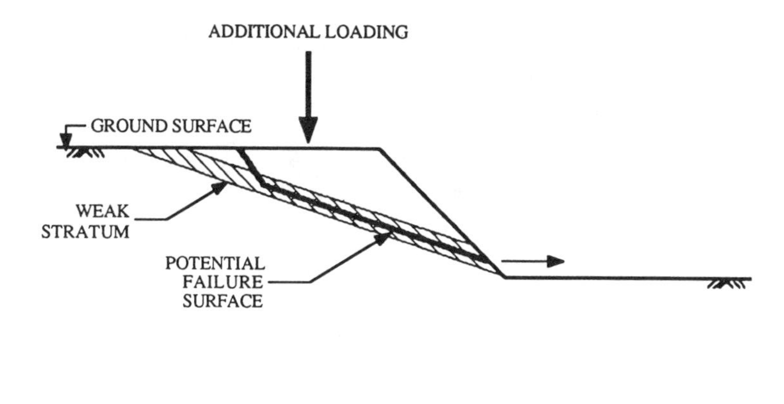

Figure 9.2 Sliding potential failure surface.

to move downward and outward. This phenomenon occurs on fill, cut, and natural slopes. The driving force of this movement is gravity. Gravity tends to slowly move the slope into a more stable position. The construction of a fill slope or a cut slope can cause this effect to increase because of the changed loading conditions at the site. Although this effect does not seem as damaging as the catastrophic slope failures discussed above, it can cause significant damage to a structure situated on top of a slope. The lateral forces that may be generated by this type of movement can damage the structure and especially any surrounding improvements such as patio slabs or parking lots.

In general, three options are available when constructing on or near a slope. The first of these options is to set the structure back far enough from the edge of the slope so the additional load of the structure will not cause the slope to fail. The second option is to construct a deep foundation system that transfers the weight of the structure to a soil or rock stratum that is beneath the zone at which failure would be likely to occur owing to the additional loading. This option, shown in Figure 9.3 will not increase the stability of the slope or arrest potential down-slope soil movement. The third option is to alter the geometry of the slope or excavate the problematic portion of the slope and reconstruct it using engineered fill soil. The latter option typically involves moving a large amount of soil and is usually only implemented on large grading projects. The remainder of this chapter discusses the

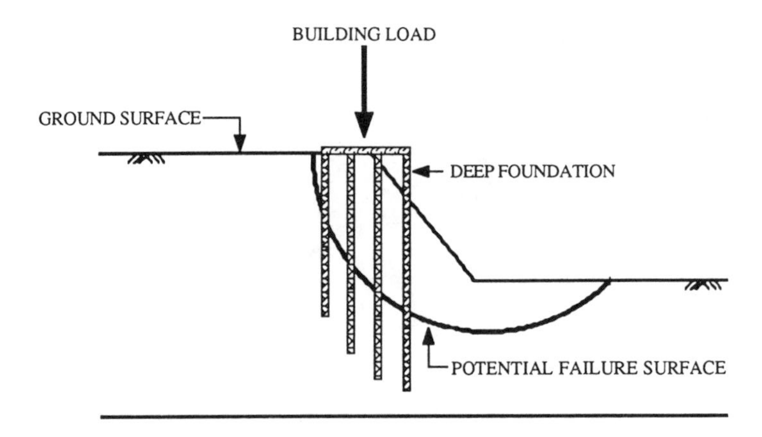

Figure 9.3 Deep foundation transferring building load to soil below potential failure zone.

first two options in more detail, as well as common design and construction oversights and difficulties when constructing on or near a slope. A fuller discussion of the third option is beyond the scope of this text; thus readers should refer to texts related to slope construction and mass grading.

SHALLOW FOUNDATIONS WITH BUILDING SETBACKS

This is undoubtedly the simplest solution. The location of the building and any surrounding improvements are moved away from the slope until their construction will no longer be detrimental to the stability of the slope and, conversely, any minor movements of the slope will not affect the structure and surrounding improvements. The design problem becomes: How far from the slope is far enough? Local or national building codes will usually specify the minimum setback distance that is required in order to use a shallow foundation system. As an example, the Uniform Building Code (ICBO, 1988) specifies a construction setback distance on top of the slope equal to at least one-third the height of the slope, but it is not required to exceed 40 feet. This configuration is depicted in Figure 9.4. This may not always be economically feasible if the lot size is too small to give up much land, especially if the setback specification includes flatwork such as walks and patios.

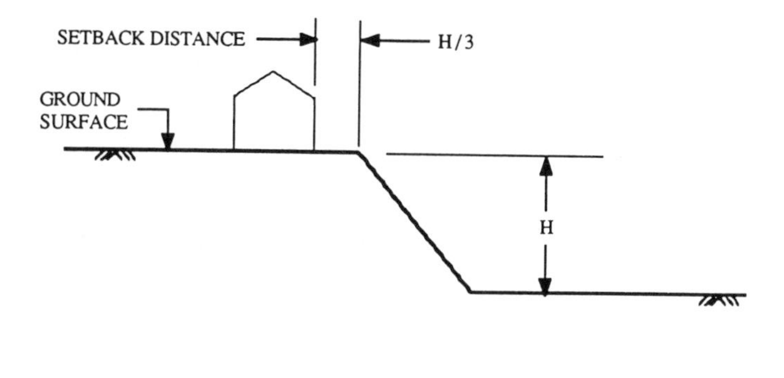

Figure 9.4 Typical building setback specification.

A setback specification simplifies the procedure, yet it may not always be the most appropriate or economical solution. For complicated geometries and geology, or when setback requirements cannot be met due to lack of space, it may be necessary to perform a detailed slope stability analysis. Computer software programs are commonly used to perform the stability analysis. Computer software packages, if properly utilized by a qualified geotechnical engineer, are capable of implementing varying geometries, soil types, groundwater levels, surface loads, and failure modes. The output consists of the locations within the slope with the lowest factors of safety against failure. If any of the potential slip surfaces exhibit a factor of safety below the minimum allowable for that region, the cause of the potential failure must be determined and remedied. For example, a change in the design of the grading plan may be required, such as changing the geometry of the slope or adding a stability fill to the face of the slope. If the failure is due strictly to the construction of the building, a deep foundation may be appropriate. If the potential slip surfaces are due to the presence of a weak layer, one option may be to excavate this layer and reconstruct the slope with engineered fill soil.

DEEP FOUNDATIONS

The purpose of a deep foundation on or near a slope is to found the structure in soil or rock below the zone that is susceptible to failure, thus placing no additional external loading on the slope. The foundation is generally stiffened with reinforcing steel to protect it against possible lateral movement. It is generally necessary to use a deep foundation system when building setback requirements for a shallow foundation cannot be met or are chosen not to be met.

To determine the depth of the foundation most local building codes specify a minimum horizontal distance from the base of a foundation to daylight (the edge of the slope) as shown in Figure 9.5. It is recommended that this specification be scrutinized for each individual site using a slope stability analysis to determine the position of the potential failure surface. With the position estimated the adequacy of the daylight specification may be assessed. Figure 9.5 illustrates a situation in which the minimum daylight specification is satisfied but the foundation does not completely penetrate the potential failure surface. This would not be a feasible solution. It would be necessary to extend the deep foundation farther into the ground than required, based strictly on the minimum daylight specification.

Recently, some local codes have been increasing the required distance to daylight. This is due to the recognition of the potential damage from slow, down-slope soil movement. The increased specification gives the foundation more embedment into soils that are more stable than those closer to the surface. The foundation will not arrest the movement, but it may be deep enough to hold the structure intact as surface soils move down slope.

Two types of foundations are commonly utilized to transfer the foundation load beyond the potential slippage zone: (1) continuous wall footings and (2) drilled shafts.

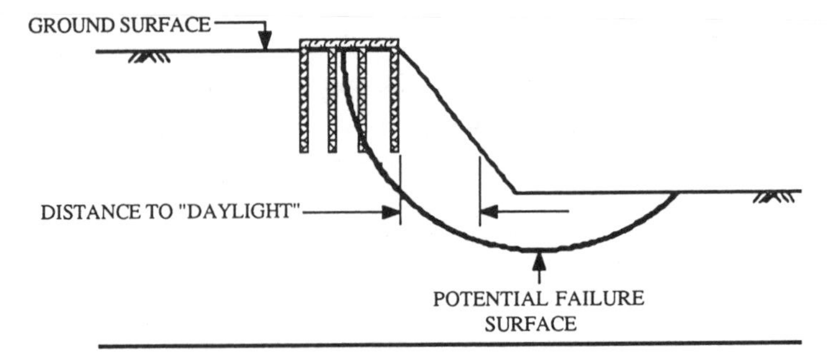

Figure 9.5 Typical minimum distance from bottom of foundation to edge of slope ("daylight").

Continuous Wall Footings

This type of foundation is fairly limited for hillside construction, because it becomes somewhat impractical to construct a continuous wall footing to a depth of much more than 4 or 5 feet. This is due to difficulties involving the placement of forms for the concrete and the placement of reinforcing steel that are compounded when working on a hillside. However, for sites that may only need several feet of embedment in order to satisfy minimum depth requirements, these foundation systems may be used effectively.

Drilled Shafts

Drilled shafts are commonly used in regions such as San Francisco Bay Area, Denver, and Southern California to support structures on or near slopes. Shaft diameters vary from 10 to 36 inches depending on the site and soil conditions and the proposed building loads (however, 30- and 36 inch-diameter shafts are relatively rare for lightly loaded structures). The depths of the shafts depend on the most likely location where the potential failure surface would occur if a shallow foundation were used. With the depth adequately determined, the drilled shaft will transfer the load of the structure to the soil below the potential failure surface. Some designs may call for a certain amount of embedment into the underlying soil stratum that is expected to provide satisfactory support. The above requirements mean that the shafts for a given structure may need to be different lengths depending on the geometry of the slope, the underlying soil conditions, and the location of the building in relation to the slope. Figure 9.6 is an example of a group of townhouses constructed on drilled shafts near the edge of a slope. The drilled shafts that can be seen in the photograph have been socketed into rock in order to support the buildings.

The amount of reinforcing steel depends on both tensile and lateral forces. Lateral forces on the shafts are derived from soil pressures and from slow downslope movement of the soil on the slope. Tensile forces are derived from wind forces on the building and the presence of any expansive soil movement. It is beyond the scope of this text to discuss methods of determining these design forces.

Occasionally an individual structure may have a combination of shallow footing foundations and end-bearing drilled shafts. This may be

Figure 9.6 Drilled shafts supporting townhouses at the top of a slope in San Diego.

an appropriate configuration when the structure is situated near the edge of the slope. The geometry of the slope, or the presence of loose soil near the edge of the slope, may be such that the portion of the house facing the slope requires drilled shafts for support, but the portion of the house facing away from the slope may only require continuous or isolated footings. This situation is illustrated in Figure 9.7. and is appropriate only for end-bearing drilled shafts. A combination of friction-supported drilled shafts and continuous or isolated footings could lead to differential settlements because of the different modes of soil support throughout the foundation system.

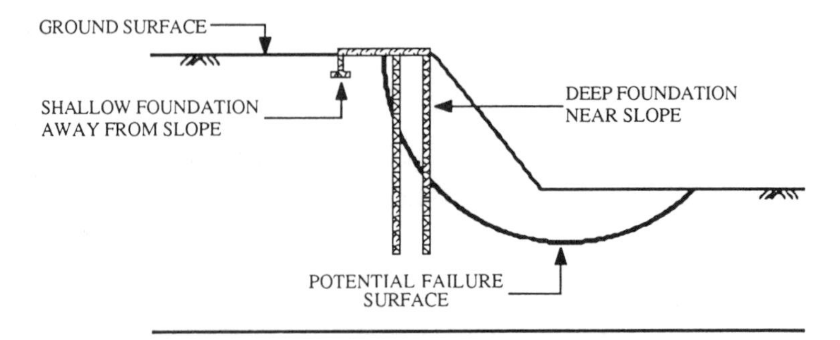

Figure 9.7 Combination foundation system.

Drilled shafts may occasionally be used to stabilize potential landslides above or below where structures are to be located. This is accomplished by drilling the shafts very close together and effectively forming a continuous wall within the slope. If properly designed and reinforced, the drilled shafts can effectively keep the hillside from sliding.

COMMON DESIGN AND CONSTRUCTION OVERSIGHTS

The presence of a hillside creates a challenging site condition for both design and construction. Additionally, the maintenance of the slope plays a key role in the performance of a structure near or on the slope. Below is a list of commonly overlooked aspects of construction on or near a hillside.

- **Same-length drilled shafts.** The length of a drilled shaft being constructed on a hillside depends on the required depth necessary to transfer the load of the structure below the potential failure surface. The depth of the potential failure surface is related to the underlying geologic and soil conditions as well as the geometry of the slope. Often the shafts are specified to penetrate a specific length into a specific competent geologic formation. The depth to this formation may vary throughout the slope. Additionally, as can be seen in Figure 9.3, the necessary length to transfer the structural load past a potential failure zone may also vary. However, the length of the shafts should be designed with an adequate factor of safety to pass sufficiently through the potential failed surface, especially since the assumed location of this surface is only an approximation.
- **Inadequate geologic and geotechnical investigation.** When a project is located on or near a slope, the geology of the site should be closely studied. The presence of any ancient landslides should be confirmed or refuted. Deep failure zones along weak layers, such as clay seams, should be investigated. Overall slope stability should also be investigated. Adhering solely to code requirements may not be adequate in all situations.
- **Equipment mobility.** Special equipment is generally required to construct structures on hillsides. Designers should be aware of the capabilities, availability, and limitations of this equipment.

Suggestions from the contractor should be given serious consideration in terms of the method of construction of the foundation system. Construction time will also undoubtedly lengthen owing to the difficulties involved in moving equipment on a slope.

- **Assuming that the foundation system will solve all potential slope failure problems.** Constructing a deep foundation will not reduce the potential for the slope to fail. The purpose of the deeper foundation is to support the building in the event of any significant movement of the slope. The slope will still need to be assessed for potential failure, especially if any fill soil is placed on top of it.
- **Influx of irrigation water.** This is more a problem of maintenance than an oversight in design or construction procedures. Before a slope is built upon, the only source of moisture is typically natural precipitation. Once a structure is erected on or near the slope, the surrounding land is undoubtedly irrigated. An increase in the water content of the soil will generally weaken the slope by reducing the shear resistance of the soil. This can lead to a catastrophic failure, or it can cause the soil on the slope to slough or move significantly downslope, causing damage to the structure at the top of the slope and potentially to any structures at the base of the slope. For areas that have a naturally arid environment this can be a common problem. For example, estimates in San Diego of the amount of irrigation water used by a typical homeowner range from 100 to 150 inches annually. This is more than 10 times the average amount of yearly precipitation normally received in the region. Saturation of the outer face of a slope that was previously relatively dry has led to numerous distress problems in this area.

  New construction may also redirect natural drainage channels in a given area, which may lead to the saturation of a slope. Careful planning must be undertaken to direct drainage water away from the slopes. Slope vegetation with deep root systems helps to anchor the soil in place and reduce the amount of movement that the soil may undergo due to over-watering. Terrace drains are commonly used to redirect surface runoff away from the slope. Homeowners should be alerted to the possible effects of over-watering and should be encouraged to landscape with drought-resistant, native vegetation. Additionally, drip irrigation systems help to regulate, isolate, and, thus, minimize the amount of irrigation water that is necessary to maintain the vegetation on or near a slope. Gray and Leiser (1982) provide recommendations for slope protection in their book *Biotechnical Slope Protection and Erosion Control.*

- **Leaking water pipes or pools.** Leaking water pipes, pools, and Jacuzzis have been known to saturate a slope, causing failure. If a leak is detected near a slope it should be repaired immediately.

SUMMARY

- Gravity is constantly attempting to force a slope into a more stable position.
- Slopes may fail due to: (1) their own weight, (2) an influx of water, which may weaken the shear resistance of the soil, (3) dynamic loading from an earthquake, (4) erosive action, (5) undercutting of the slope due to construction, and (6) the addition of weight at the top of a slope due to construction.
- Construction of a structure on or near a slope can increase the probability of a slope failure.
- There are two general types of slope failures: rotational and sliding.
- A slow downward movement of a slope, commonly referred to as *creep,* affects construction on or near a slope by creating significant lateral forces.
- Setting a structure a specified distance from a slope and constructing a deep foundation system are two commonly used options when faced with construction near a slope.
- It is recommended that the adequacy of a setback specification be assessed for each site's unique geometry.
- Depths of deep foundations should be designed based on the potential failure surface location and the geometry of the slope. Deep foundations of different length should be anticipated.
- Combinations of continuous or isolated spread footings with end-bearing drilled shafts are appropriate for near-slope loose-soil conditions.
- Overall slope stability and weak layers should be adequately assessed for hilly sites.
- Costs typically increase for construction on or near a slope because of the difficult access and limited mobility for the equipment.

- The purpose of a deep foundation near a slope is to transfer the load of the structure beyond the potential failure zone.
- The influx of moisture to the slope (commonly due to landscape irrigation or leaking pipes or pools) can greatly affect the stability of the slope.
- Erosion of portions of the slope may affect the stability of the slope.

RELATED REFERENCES

ADSC: The International Association of Foundation Drilling and DFI: Deep Foundation Institute, *Drilled Shaft Inspector's Manual,* 1st Edition, ADSC, Dallas, 1989.

BOWLES, J. E. *Engineering Properties of Soils and Their Measurement,* 3rd Edition, McGraw-Hill Book Co., New York, 1986.

CHOWDHURY, R. N. *Slope Analysis,* Elsevier Scientific Publishing Co. Amsterdam, The Netherlands, 1982.

COLE, K. W., *Foundations,* Thomas Telford Ltd., London, 1988.

DAS, B. M., *Principles of Foundation Engineering,* PWS Publishers, Boston, 1984.

FLEMING, W. G. K., ET AL., *Piling Engineering,* John Wiley & Sons, New York, 1985.

FRENCH, S. E., *Introduction to Soil Mechanics and Shallow Foundations Design,* Prentice Hall, Englewood Cliffs, NJ, 1989.

GRAY, D. H, AND LEISER, A. T., *Biotechnical Slope Protection and Erosion Control,* Van Nostrand Reinhold, New York, 1982.

GREER, D. M., AND GARDNER, W. S., *Construction of Drilled Pier Foundations,* John Wiley & Sons, New York, 1986.

HOLTZ, R. D., AND KOVACS, W. D., *An Introduction to Geotechnical Engineering,* Prentice Hall, Englewood Cliffs, NJ, 1981.

INTERNATIONAL CONFERENCE OF BUILDING OFFICIALS, *Uniform Building Code,* ICBO, Whittier, CA, 1988.

LA ROCHELLE, P., "Problems of Stability: Progress and Hopes," *The Art and Science of Geotechnical Engineering at the Dawn of the Twenty-First Century,* Prentice Hall, Englewood Cliffs, NJ, 1989.

LAMBE, T. W., AND WHITMAN, R. V., *Soil Mechanics,* John Wiley & Sons, New York, 1969.

LAMBE, T. W., SILVA, F., AND LAMBE, P. C., "Expressing the Level of Stability of a Slope," *The Art and Science of Geotechnical Engineering At the Dawn of the Twenty-First Century,* Prentice Hall, Englewood Cliffs, NJ, 1989.

OLSON, R. E., AND LONG, J. H., "Axial Load Capacity of Tapered Piles," *The Art and Science of Geotechnical Engineering At the Dawn of the Twenty-First Century,* Prentice Hall, Englewood Cliffs, NJ, 1989.

PECK, R. B., "Stability of Natural Slopes," *Journal of the Soil Mechanics and Foundation Division,* ASCE, 93, SM4, pp. 403–417, 1967.

REESE, L. C., AND O'NEILL, M. W., *Criteria for the Design of Axially Loaded Drilled Shafts,* Center for Highway Research, Univ. of Texas, Austin, August 1971.

REESE, L. C., AND O'NEILL, M. W., *Drilled Shafts: Construction Procedures and Design Methods,* for the U.S. Dept. of Transportation, FHWA-HI-88-042, ADSC: The International Association of Foundation Drilling, ADSC-TL-4, Dallas, 1988.

SMITH, G. N., *Elements of Soil Mechanics,* 6th Edition, BSP Professional Books, Oxford, 1990.

TAYLOR, D. L., *Low Capacity Pier Foundations, Denver Metropolitan Area.* Thesis presented to the University of Colorado, Denver, in partial fulfillment of the requirements for the degree of Master of Engineering, 1990.

TERZAGHI, K., *Theoretical Soil Mechanics,* John Wiley & Sons, New York, 1943.

TOMLINSON, M. J., *Pile Design and Construction Practice,* 3rd Edition, Palladian Publications, London, 1987.

ZEEVAERT, L., *Foundation Engineering for Difficult Subsoil Conditions,* 2nd Edition, Van Nostrand Reinhold, New York, 1983.

CHAPTER 10

Foundation Construction Quality Control and Maintenance

INTRODUCTION

Conversations with engineers and contractors from around the country reveal that foundation quality control is an important issue in most regions, especially when challenging soil and site conditions are encountered. One comment heard during these conversations aptly sums up the importance of quality control: "The best drilled shaft in the world will not do any good if it is designed [or constructed] in the wrong place [within the structure]." This is true for any foundation system.

Foundation quality control includes much more than just ensuring that the foundation system is properly located. Quality control begins with the geotechnical investigation of the site and continues through the final placement of the concrete. Additionally, once the structure is completed, it becomes the owner's responsibility to maintain the foundation system throughout the life of the structure.

One way to stress the importance of quality control is to illustrate what can happen without it. Chapter 4 included a case history of a distressed home constructed on expansive clay. To refresh the reader's memory, this house was built on a drilled shaft foundation system. Three aspects of the project that were critical to the success of the foundation system were overlooked. First, a geotechnical investigation was not

> **Quality control begins with the geotechnical investigation of the site and continues through the final placement of the concrete. Additionally, once the structure is completed, it becomes the owner's responsibility to maintain the foundation system throughout the life of the structure.**

performed and, thus, the expansiveness of the soil was not detected. Second, the designer was unfamiliar with the design of drilled shafts in expansive soils and designed shafts of insufficient length for the expansive soil conditions. The foundation system functioned like an isolated spread-footing foundation with round footings—which, as discussed in Chapter 4, are rarely appropriate for highly expansive soils.

The third aspect leading to the foundation failure was the lack of qualified inspection of the foundation system during construction. This led to an overpouring of the concrete for the shafts. As discussed in Chapter 4, this condition is referred to as "mushrooming" and is detrimental because a horizontal surface is formed directly on top of the soil, which the expansive clay can push up against.

This example illustrates what may occur without quality control. The designer should have consulted with someone who was familiar with drilled shaft design in the region. Presumably the designer knew that drilled shafts were commonly used in the region for construction on expansive soil but wasn't familiar with how they performed. The designer knew just enough to be dangerous. Inspection during the concrete pour might have eliminated the mushrooming of the shafts. Additionally, if an inspector had an appreciable amount of experience in the region, this individual might have questioned the depth of the shafts.

The effectiveness of quality control for foundations is typically based on two factors: money and experience. Builders constantly monitor their costs. One of the first phases of a project where costs may be minimized, with unanticipated results, is the geotechnical investigation. Many contractors and engineers from around the country who were interviewed in the preparation of this text pointed to insufficient soil investigations as the ultimate cause of many foundation failures. Perhaps even more common is the occurrence of unexpected conditions that arise during construction and are not reevaluated by someone qualified to do so. When unexpected conditions arise during construction, the builder is faced with spending more money by calling a qualified geotechnical engineer out to the site. Additional costs due to the delay in construction are likely. These costs must be weighed against the risk of incurring additional costs in the future due to a foundation distress problem, as well as the risk to the builder's reputation.

> The effectiveness of quality control for foundations is typically based on two factors: money and experience.

Some localities do not require a geotechnical investigation and/or observation during construction other than a periodic site visit by the building department inspector. This creates another situation in which the builder must decide whether to hire an engineer to perform these functions. Unfortunately, many builders will commission only the minimum amount of quality control required by the local building codes and/or building insurance companies.

> One of the first phases of a project where costs may be minimized, with unanticipated results, is the geotechnical investigation. Many contractors and engineers from around the country who were interviewed in the preparation of this text pointed to insufficient soil investigations as the ultimate cause of many foundation failures.

The experience of those performing observation services is also a factor in the adequacy of quality control. More often than not, the building inspector or the on-site soil technician or geotechnical engineer does not have a great deal of experience. The construction site then becomes a training ground. An additional problem is that observing personnel may be qualified to inspect one aspect of the foundation system but not all aspects. For example, a geotechnical engineer may be hired to observe a post-tensioned slab installation. The engineer may be qualified to observe the preparation of the subgrade yet may not have any experience with post-tensioned systems.

What is the solution to this multifaceted problem? Full-time observation would price many builders and potential homeowners out of the market. Additionally, it would not solve the problem of the inexperienced inspector. Requiring an experienced engineer to be at the site full time would be costly. Therefore, the builders must be informed of the importance of quality control, and guidelines for quality control should be provided to technicians and engineers.

Before continuing, it is important to stress the importance of safety precautions that should be adhered to at any construction site. Safety precautions related to foundation construction are typically related to the excavations of the foundations themselves. It is important to remember that all federal Occupational Safety and Health Administration (OSHA) standards should be adhered to at any construction site and specifically as they relate to deep excavations, drilled shafts, and pile driving. Safety should be foremost in every individual's mind at the job site.

The remainder of this chapter presents some guidelines for foundation observations of different foundation systems. Individual sites may require more stringent quality control procedures than those presented here. The level of quality control should be discussed and agreed upon between the builder and the engineer, preferably before construction begins. The chapter concludes with a discussion of professional interaction and some suggestions for building owners regarding foundation maintenance.

GUIDELINES FOR FOUNDATION CONSTRUCTION QUALITY CONTROL

The purpose of this section is to present some general guidelines for foundation observations and quality control for different lightly loaded foundation systems. It is beyond the scope of this guide to present a comprehensive checklist pertaining to each foundation system, cross referenced with each challenging soil or site condition. Readers who will be performing inspections professionally are encouraged to compile a checklist specifically for their region. A detailed example of a checklist for expansive soil conditions is presented in Appendix A.

Site Investigation

One of the most important items to remember when conducting a geotechnical investigation is that no two sites are the same. Because of this, it is impossible to suggest a uniform sampling schedule that would be pertinent and practical for all sides. However, there are some general guidelines that may be followed. Experience, or talking with people experienced with *similar* sites is highly recommended.

Once the field investigation has begun, it is important to keep in mind how the proposed project will affect the site. For example, will there be any significant cuts or fills on the site; what magnitude of loading is expected; and, to what depth will the soil be affected by the construction? Other items to keep in mind include potential geologic hazards such as earthquake faults or potential landslides. It is also important that an engineer with considerable experience in the region visit the site and be involved in the investigation and scheduling of the laboratory testing.

If there are any known regional conditions that require special foundations, such as those described in earlier chapters of this text, the investigations should carefully scrutinize the potential for these conditions to exist on site. For example, a regional presence of expansive soil would necessitate a laboratory testing program that would evaluate the swell potential of the on-site soils. Another example would be the excavation of test trenches in an area that may be susceptible to limestone solution voids or sinkhole formation.

Another helpful aid in an investigation is a review of any existing aerial photographs. Volumes of air photographs can be purchased from aerial photograph companies in most large cities. Photographs often reveal undocumented site gradings performed in the past.

Additionally, historic air photos might reveal that the previous use of a site poses environmental hazards today. It is beyond the scope of this book to discuss environmental concerns for a site; however, if it is suspected that environmental hazards do exist at a site, an environmental engineering company should be retained to conduct a preliminary site assessment. In addition, if any suspected hazardous material or abnormally discolored soil is encountered during the geotechnical investigation or construction, the project should be temporarily shut down until an environmental engineering company has assessed the extent of the hazardous material and has provided recommendations pertaining to its safe removal.

The following is a brief list of questions that may be helpful when planning and performing a geotechnical investigation.

1. How many borings will adequately reveal the existing soil conditions, taking known regional conditions into account?
2. Will the excavation of exploratory trenches significantly increase the amount of information or adequately replace information to be obtained from soil borings?

3. Have aerial photographs been reviewed?
4. Have previous soil reports of nearby sites been reviewed?
5. What is the extent of the proposed site grading? Will there be large cuts and fills? Are cut/fill transitions located at proposed building pads?
6. How deep should the borings be? To what depth will the anticipated foundation system affect the soil? How compressible are the in situ soils? Are large settlements anticipated?
7. Are there any known regional conditions, such as expansive soils, slope stability problems, highly compressible soils, etc., that may require a special foundation?
8. What methods of soil sampling will be required in the field so that pertinent laboratory tests may be conducted? For example, if it is anticipated that the soil is highly compressible, have undisturbed samples been obtained to perform consolidation tests? Similarly, if there is a possibility that some of the soil will need to be removed and recompacted, have representative bulk samples been obtained in order to conduct a compaction test?
9. Is there a relatively shallow groundwater table or will site drainage present a problem?
10. Will special equipment be necessary to perform the investigation because of site accessibility problems such as hillsides or soft surface soils?
11. Are there any geologic hazards that will significantly affect the site such as faults and regional seismicity, or potential landslides and slope stability problems?
12. Has the local availability of specialized construction equipment been considered when recommendations are being made? For example, for a hillside drilled-shaft installation, what is the availability of a hillside drill rig? Another example is if dynamic compaction is being considered, is there a company within a reasonable distance that provides this service?

Foundation Plan Review and Construction Control

Continuity of a project from beginning to end helps to maintain high levels of quality control. For this reason, it is important that the project plans be reviewed by the same geotechnical engineer who made the original foundation recommendations. It is also generally desirable to have the same company monitor the geotechnical aspects of construction for the project.

1. Has a geotechnical engineer reviewed the foundation plans to confirm that the original foundation recommendations have been implemented?
2. Has a geotechnical engineering company (preferably the same company that prepared the soil investigation) been retained for quality-control monitoring during project construction?

Continuous Footings and Isolated Spread Footings

Below is an abbreviated checklist of questions that will facilitate the inspection of continuous wall-footing or isolated spread-footing excavations.

1. Are the soil conditions in the excavation similar to those that were described in the geotechnical investigation report?
2. Is the bearing capacity of the soil adequate to support the foundation load?
3. Are the dimensions, such as the depth below adjacent finish grade, or the width of the footings in general conformance with the plans and the geotechnical report?
4. Is the amount of steel reinforcement in general conformance with the structural plans and the geotechnical report? Is the steel reinforcement properly located within the footing excavation? Is the reinforcement free of corrosion other than surface rust?
5. Are there any loose materials in the excavations such as rocks, trash, or soil that have fallen or sloughed into the footing bottoms?
6. If the soil conditions in the bottom of the footing are clayey, are shrinkage cracks beginning to form due to moisture loss in the soil? If so, the bottom of the excavation should be moisture conditioned until the cracks close.
7. Is the footing excavation properly moisture conditioned to accept concrete? The footing excavation should not have standing water in it that will weaken the concrete. Dry soil conditions in the bottom of the excavation will draw moisture from the concrete and cause it to cure too rapidly. Rapid curing will lead to excess cracking in the footing.

8. Is the ground frozen or is the ambient temperature too low? (See Chapter 7).
9. Is there any seepage of groundwater into the footing excavation?

Conventionally Reinforced Slabs

Below is an abbreviated checklist of questions that will facilitate the inspection of conventionally reinforced slab-on-grade foundations.

1. Has the subgrade been properly compacted, moisture conditioned, and cleaned of any loose material and debris?
2. If thickened edges and/or deepened beams are planned, have they been excavated in accordance with the plans and are the bearing soils similar to those described in the geotechnical report?
3. Has a capillary break been used, such as a layer of gravel or sand, plus a vapor barrier? This is not required, but it is generally recommended. Additionally, a sand layer offers a relatively uniform surface upon which to place the concrete and aids the uniform curing of the slab by providing a surface beneath the slab for water to hydrate from the concrete.
4. If steel reinforcing bars are being used, are they in general conformance with the structural plans and the recommendations of the geotechnical engineer, and are they free from corrosion other than surface rust?
5. If welded wire mesh is being used, is it being properly lifted into the slab or is it being forced down to the bottom of the slab while the concrete is being placed?
6. Is the site grading sloped away from the slab on all sides of the structure?

Post-Tensioned Reinforced Slabs

Below is an abbreviated checklist of questions that will facilitate the inspection of post-tensioned reinforced slab-on-grade foundations. One of the arguments against the post-tensioned system is that it requires a much more rigorous quality-control program. The inspection items listed above for the conventionally reinforced slabs are relevant for the post-tensioned system, as well as those listed below. The list below is an example of some of the key items that should be inspected for a post-tensioned system. A complete list may be found in *Field Procedures Manual for Unbonded Single Strand Tendons* (PTI, 1989a).

1. Are fixed end wedges evenly and adequately seated in the anchorage?
2. Is the plastic sheathing of sufficient and uniform thickness?
3. Does the strand appear to be free of corrosion (other than surface rust) where sheathing and grease are removed at stressing ends?
4. Are the tendon profiles smooth and correctly shaped (parabolic, circular, or straight without localized reverse curves) between reference points?
5. Do the tendons have localized horizontal wobble?
6. Has the method of concrete placement been reviewed as to its effect on tendon stability during placement? Concrete vibrators should not come in contact with the tendons.
7. Has the concrete reached proper strength according to the plans before stressing of the tendons is performed?
8. Are the tendons perpendicular to the anchorage and is the anchorage parallel to the face of the concrete? Are the wedges seated evenly?
9. Are the grippers and stressing jack free of soil, plastic, and cement paste?
10. Are the tendon tails cut off well inside the pocket to allow proper grout cover?

Readers who may be involved in the inspection of post-tensioned systems are again encouraged to obtain a copy of *Field Procedures Manual for Unbonded Single Strand Tendons* (PTI, 1989a).

Drilled Shaft and Grade Beam Systems

Below is an abbreviated checklist of questions that will facilitate the inspection of drilled shaft and grade beam systems. It is intended to give the reader an overview of the necessary items involved in the quality control for drilled shaft construction. Readers are encouraged to consult the *Drilled Shaft Inspector's Manual* (ADSC, 1989) for additional inspection details. Complete construction procedures for the installation of temporary casing and concrete tremie have been omitted because these procedures are relatively rare for lightly loaded construction projects.

1. Are the shafts properly located as shown on the foundation plan? Typically, the shaft locations will have been previously surveyed; however, the inspector should be familiar with the foundation plan and be able to recognize any gross errors in shaft location.
2. Is the shaft being drilled relatively plumb? The shaft should be within 2% of plumb over its total length (ADSC, 1989).
3. Has the shaft been drilled to a proper bearing stratum or to the design evaluation?
4. Is the bottom of the shaft relatively free of loose material? The allowable amount of loose material in the bottom depends on whether the drilled shaft is designed for end-bearing or side friction, the characteristics of the bearing stratum and the tolerable differential settlement of the building. Typically, the allowable amount of loose material will be in the specifications; otherwise the material must be sufficiently excavated to the satisfaction of the inspector.
5. If shear rings are being used, has the loose material created from their excavation been removed from the bottom of the shaft?
6. Is there any groundwater in or actively seeping into the shafts? Generally, this condition will be anticipated from the geotechnical investigation, and the contractor will be prepared to tremie the concrete. One to two inches of water at the base of the shaft is acceptable at the time of the concrete pour; otherwise it will be necessary to cease operations until a tremie pour is available. If the rate of seepage into the shaft is low, the water can be pumped out of the shaft and the concrete can be poured once the water has been removed. A tremie pour consists of pumping the concrete through a pipe to the bottom of the shaft. As the shaft fills with concrete, the water is displaced and flows out of the top of the shaft.
7. Are the shafts being covered after drilling until they are ready to be filled with concrete? This eliminates both a safety hazard and the potential for loose surface materials to fall into the shaft.
8. Has the drilled shaft been left open overnight prior to the concrete pour? If so, the shaft should be covered and the bottom of the shaft should be checked again for cleanliness just prior to the steel and concrete placement to ensure that the shaft has not experienced any caving or sloughing. If caving or sloughing has occurred then it is necessary to remove the loose material from the bottom of the shaft.

9. If a belled shaft is required, are the dimensions of the bell in conformance with the plans? Has the loose material generated from the bell excavation been cleaned out of the shaft?
10. If reinforcing steel is required, is the steel reinforcement in general conformance with the foundation plan for each drilled shaft? Is the steel free from corrosion other than surface rust?
11. Is the concrete designed with a relatively high slump? Slump designs should be a minimum of 5 inches to ensure an adequate concrete/soil bond and to avoid honeycombing of the concrete.
12. Has all excess concrete been removed from the ground surface to eliminate mushrooming of the shafts?
13. If void boxes are being used beneath the grade beams, has their integrity been maintained up to the time of concrete placement?

Driven Pile Systems

Below is an abbreviated checklist of questions that will facilitate the inspection of driven pile foundations. Local building codes contain details on the necessary requirements for pile installation.

1. Do the piles delivered on-site meet the specifications of the plans? Items that should be checked include the diameter (or other applicable dimensions if the piles are not designed with a circular cross section), the length, and the taper amount (if any).
2. Has the integrity of the piles been checked?

   Timber piles should be checked for straightness, the twist of the grain, trimming of knots and their frequency, holes, cracks, and bark removal.

   Precast concrete piles should be checked for straightness, cracks or chips, length of cure at the time of installation, concrete strength at the time of transport to the site, and of installation based on cylinder breaks.

   Steel piles should be checked for straightness, corrosion, and integrity of any welds.
3. Are the piles properly located as shown on the foundation plan? Typically, the pile locations will have been previously surveyed; however, the inspector should be familiar with the foundation plan and be able to recognize any gross errors in pile location.

4. Are protective cushions being used during driving for precast concrete piles?
5. If the pile locations are being predrilled, are the predrilling depths shallow enough to leave an adequate thickness of undisturbed soil into which to drive the piles?
6. Are splicing of piles permitted according to the local building code?
7. Have refusal driving rates been determined before driving has begun? This is important in order to avoid damage to the piles from overdriving.
8. Are the piles being installed plumb?
9. Have the proper design capacities been achieved based on driving rates and hammer energies?
10. Have minimum design tip elevations been achieved?
11. Were the piles excessively damaged during driving?
12. Is the hammer capable of supplying the minimum energy as specified in the plans and/or local codes?

Grading Observation and Testing

Below is an abbreviated checklist of questions that will facilitate the observation of grading and backfill operations. For additional information readers are encouraged to refer to *Construction Of and On Compacted Fills* (Monahan, 1986).

1. Is the site free of shrubs, weeds, and debris prior to the placement of fill soils and/or the construction of the foundation?
2. Is the fill material of consistent and clean nature, free of trash, debris, and organic material?
3. Is the appropriate compaction equipment being used relative to the soil type being compacted?
4. Are the materials being compacted expansive in nature? If so, has this been taken into account with regard to the percent relative compaction and/or the moisture content?
5. Are the compaction lifts too thick?
6. Is the required compactive effort being achieved as determined by the dry density? Are the moisture levels within design specifications? If not, is the contractor correcting the problem(s)?

7. How often is the compacted material being tested to see whether the placement densities are within the specifications? Is this within the geotechnical engineer's recommendations?
8. What method is being used to test the compaction? Is it an appropriate method for the given soil conditions? Is it in accordance with the geotechnical engineer's recommendations? Is a rock correction necessary? (See Chapter 6.)
9. Is the soil that is being compacted frozen?

COMMUNICATION AND QUALITY CONTROL

During the construction of a project, two things are required in order to maintain a high level of quality control: first, a preconstruction meeting and, second, field communication during construction. Although this sounds simple, both procedures are quite often not carried out. The results can be potentially catastrophic and, at a minimum, lead to construction delays.

During the construction of a project, two things are required in order to maintain a high level of quality control: first, a preconstruction meeting and, second, field communication during construction.

The preconstruction meeting is important because it confirms the methodology and order of construction anticipated by all of the involved parties. Engineers from around the country stress the fact that although they typically perceive preconstruction meetings as unimportant, there is always something discussed at the meeting that one of the parties was previously unaware of.

An excellent case history in which a preconstruction meeting was not conducted, resulting in problems, involves a site with a loose, compressible layer of soil that was scheduled for improvement via the technique of dynamic compaction. The grading plan revealed that the site would need additional fill to achieve finish grade. The contractor misunderstood the order in which the dynamic compaction and the placement of the additional fill were to be conducted. The fill was placed first and the contractor then proceeded with the dynamic compaction. The intent of the dynamic compaction, however, was to densify the existing native soils, not to compact the fill soils. Once the fill soils had been placed, the dynamic compaction had much less effect on the underlying native soils. It was necessary to remove all the fill soils and proceed with the dynamic compaction procedure a second time. The extra cost and subsequent delay could have easily been avoided had a preconstruction meeting been held.

Another scenario that commonly occurs is a lack of communication in the field between contractors and engineers. This is generally not a problem until changed conditions occur. The problems become magnified when the contractor and/or developer does not request the presence of the engineer as soon as conditions change. Once the engineer does become involved, it may be too late to avoid costly construction delays.

FOUNDATION MAINTENANCE

As with any product, a foundation will perform better and last longer if it is properly maintained. There are three general areas of foundation maintenance that are all related. They are: (1) the restraint of irrigation water; (2) site grading, and (3) landscaping. In essence, each of these is more directly related to the condition of the subgrade soils on which the foundation is bearing rather than the foundation itself. Each of these is specifically related in some way to the moisture conditions of the bearing soil beneath the foundation.

Irrigation Water

The control of irrigation water is very important, especially in arid environments. Saturation of the subgrade soils will weaken their bearing strength. This may lead to differential settlements if certain areas surrounding a building are constantly being irrigated. If the subgrade soils are expansive in nature, the influx of water may cause the soil to expand and subsequently place the building in distress. If the site is on or near a slope, the influx of water increases the potential for a localized foundation failure or a failure of the entire slope.

One way to control irrigation is by installing a drip-irrigation system. This system regulates and isolates the amount of water that enters the soil. If the system is properly designed, the amount of water that enters the soil should be enough to maintain the landscaping, yet not saturate the soil underlying the root systems. Another way to control or restrain the irrigation water is to manipulate the site grading and drainage, as well as the landscaping, in such a manner that irrigation water does not affect the foundation system. The latter two methods are discussed in the following sections.

Site Grading and Drainage

The concept of providing adequate drainage is simple. Constant, positive drainage away from the structure should be maintained on all sides of the structure at all times. For example, the Colorado Geological Survey (1987) recommends a minimum slope away from a structure of 1% for bare or paved areas and a minimum of 5% for landscaped areas within the first 10 feet of the foundation system. This survey states that the preferred slope within 10 feet of the foundation is between 10% and 15%. If the slope is much larger than 15%, surface runoff may cause the slope to erode.

If it is critical that water be kept away from the foundation and the bearing soils because of the volatility of the soils and/or the sensitivity of the structure to foundation movements, consideration should be given to the installation of subdrains around the perimeter of the building. The purpose of the subdrains is to channel water that has seeped into the soil away from the foundation. The drain is typically constructed to outfall into the gutter or storm drain system. An example of a typical subdrain around a house foundation is depicted in Figure 10.1. The subdrain trench is generally backfilled with gravel to create a conduit for the water to travel through to facilitate its entrance to the perforated drain pipe. It is essential that impermeable waterproofing material is placed on the structure side of the trench to keep water in the trench from soaking into the subgrade soils. Roof runoff should also be controlled in roof drains with a downspout that drains directly onto a splashblock, as shown in Figure 4.14, or preferably into a closed pipe

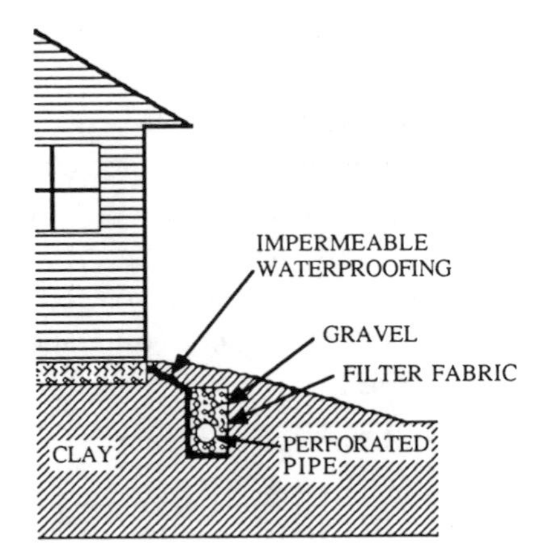

Figure 10.1 Typical perimeter subdrain.

system that channels the water away from the foundation and into the storm drain system.

It is necessary to maintain the slope and drainage system throughout the life of the structure. It is possible for the backfill material adjacent to the structure to settle with time. This settlement will reduce the positive drainage away from the structure. The area should be compacted by hand as much as possible and additional fill should be brought in to rebuild the drainage swale. Drainage structures should be checked periodically for clogs or deterioration.

Landscaping

There are two aspects to consider when landscaping with respect to maintaining a foundation system: (1) the location of landscaping structures and vegetation, and (2) the choice of vegetation. As mentioned above, the decisions about landscaping are related to water drainage, and the question "Where will the water go if this is placed here?" should always be kept in mind.

Any landscaping structures such as patios and walks, flower beds and walls, should be constructed in such a manner that water does not pond adjacent to the foundation system. It is not recommended that flower beds or raised planters be placed adjacent to the structure because they will require irrigation water that will potentially drain into the subgrade soil below.

The placement of the vegetation is also a factor. Shrubs with deep roots systems and trees can affect the moisture content of the subsurface soils. Because the root systems are far-reaching, they can affect the soil directly beneath the foundation. Changes in the moisture level may cause the soil to change volume and place the foundation in distress. A good rule of thumb is to plant trees and large shrubs a horizontal distance away from the foundation equal to the predicted ultimate height of the plant or tree, as shown in Figure 10.2.

The type of vegetation planted may also help maintain the foundation. Plants need varying amounts of water to survive. If it is absolutely

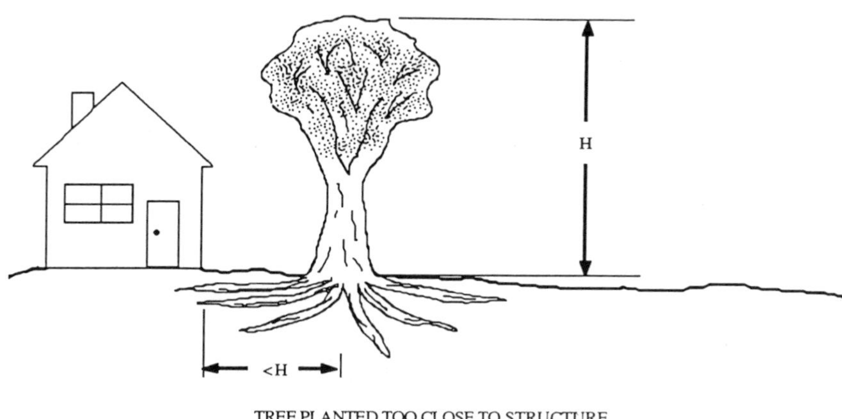

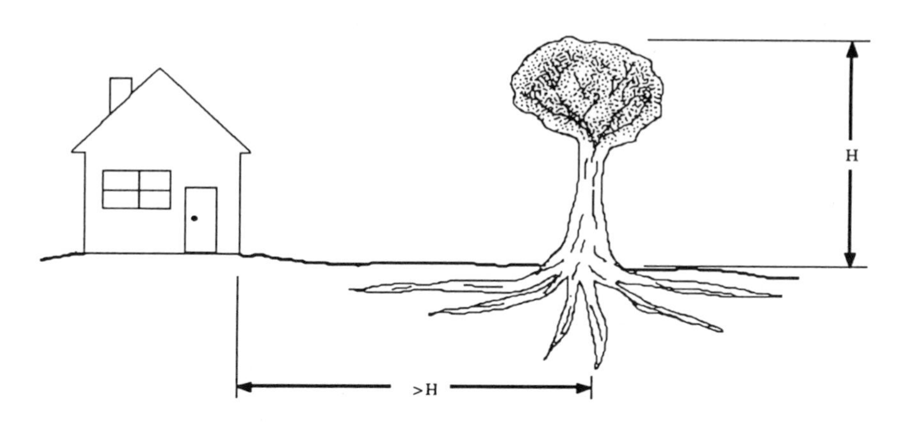

Figure 10.2 Tree planted too close to structure undermining foundation and tree planted far enough from structure to not affect the foundation.

necessary to plant near the structure, choose plants that do not require large quantities of water so as to help maintain the foundation. *Taylor's Guide to Water-Saving Gardening* (1990) offers a selection of attractive landscaping and low water-demanding vegetation to aid in choosing this type of landscaping.

RELATED REFERENCES

ADSC: The International Association of Foundation Drilling and DFI: Deep Foundation Institute, *Drilled Shaft Inspector's Manual,* 1st Edition, ADSC, Dallas, 1989.

American Concrete Institute, *Guide to Residential Cast-in-Place Concrete Construction,* Report No. ACI 332R-84, American Concrete Institute, Detroit, 1989.

Brown, R. W., *Residential Foundations—Design, Behavior and Repair,* 2nd Edition, Van Nostrand Reinhold, New York, 1984.

Building Research Advisory Board of the National Research Council, *Criteria for Selection and Design of Residential Slabs-on-Ground,* Report No. 33 to the Federal Housing Administration, National Academy of Sciences, Washington, DC, 1968.

Colorado Geological Survey, *Home Landscaping and Maintenance on Swelling Soil,* 1st Revision, Special Publication 14, Department of Natural Resources, Denver, 1987.

Das, B. M., *Principles of Foundation Engineering,* PWS Publishers, Boston, 1984.

DYWIDAG Systems International, *DYWIDAG Monostrand Posttensioning System,* 1990.

Fleming, W. G. K., et al., *Piling Engineering,* John Wiley & Sons, New York, 1985.

French, S. E., *Introduction to Soil Mechanics and Shallow Foundations Design,* Prentice Hall, Englewood Cliffs, NJ, 1989.

Gray, D. H., and Leiser, A. T., *Biotechnical Slope Protection and Erosion Control,* Van Nostrand Reinhold, New York, 1982.

Greer, D. M., and Gardner, W. S., *Construction of Drilled Pier Foundations,* John Wiley & Sons, New York, 1986.

Holtz, R. D., and Kovacs, W. D., *An Introduction to Geotechnical Engineering,* Prentice Hall, Englewood Cliffs, NJ, 1981.

Monahan, E. J., *Construction Of and On Compacted Fills,* John Wiley & Sons, New York, 1986.

NAHB RESEARCH FOUNDATION, INC., *Residential Concrete,* National Association of Home Builders, Washington, DC, 1983.

POST-TENSIONING INSTITUTE, *Field Procedures Manual for Unbonded Single Strand Tendons,* Post-Tensioning Institute, Phoenix, AZ, 1989(a).

POST-TENSIONING INSTITUTE, *Design and Construction of Post-Tensioned Slabs-on-Ground,* 1st Edition, Post-Tensioning Institute, Phoenix, AZ, 1989(b).

REESE, L. C., AND O'NEILL, M. W., *Criteria for the Design of Axially Loaded Drilled Shafts,* Center for Highway Research, Univ. of Texas, Austin, August 1971.

REESE, L. C., AND O'NEILL, M. W., *Drilled Shafts: Construction Procedures and Design Methods,* for the U.S. Dept. of Transportation, FHWA-HI-88-042, ADSC: The International Association of Foundation Drilling, ADSC-TL-4, Dallas, 1988.

SCHROEDER, W. L., *Soils in Construction,* John Wiley & Sons, New York, 1984.

SEED, H. B., "A Modern Approach to Soil Compaction," *Proceedings of the California Street & Highway Conference,* 11th Proceedings, 1959, pp. 77–93.

SMITH, G. N., *Elements of Soil Mechanics,* 6th Edition, BSP Professional Books, Oxford, 1990.

Taylor's Guide to Water-Saving Gardening, Houghton Mifflin, Boston, 1990.

TOMLINSON, M. J., *Pile Design and Construction Practice,* 3rd Edition, Palladian Publications, London, 1987.

CHAPTER 11

Foundation Innovations and Conclusions

INTRODUCTION

The purpose of this chapter is to discuss some of the recent innovations in the lightly loaded foundation industry. Also included at the end of this chapter are closing comments regarding lightly loaded foundations at sites with challenging soil and site conditions. The innovations are presented in order to give readers ideas that might improve the quality of the lightly loaded foundation industry in their region.

Innovations enhance the lightly loaded foundation industry by creating more effective and less costly foundation alternatives. Innovative systems are particularly applicable to challenging site conditions because they may offer a solution where a conventional system may not. Many of the innovative foundation systems resulted from the need for a more effective foundation solution for a particular challenging site condition.

There are few unique and innovative foundation systems entering the market today. The majority of the innovations are modifications of existing systems. The reasons for this lack of innovation are related to cost and resistance to change. Innovations typically take money to develop. In addition, they may also cost more money up front to construct, with the savings coming from reduced maintenance and repair costs.

Even if a builder is willing to spend the extra money and accepts the risks involved with implementing a new system, the project is often thwarted by regulatory agencies that are not willing to approve something with which they are unfamiliar.

The following innovations are briefly presented to illustrate that alternatives or modifications to the more traditional foundation systems are being used in certain regions for challenging soil and site conditions. The astute reader will realize that many innovations discussed here are related to expansive soils. This is because the presence of expansive soil is one of the more challenging site conditions and thus has encouraged designers to pursue new alternatives.

EXAMPLES OF INNOVATIVE SYSTEMS

Precast Grade Beam System

A manufacturer and concrete company in northern California designs and constructs precast grade beams for drilled shaft and grade beam foundations. The system is very similar to a traditional drilled shaft and grade beam system except that the grade beams are precast at the manufacturing plant and are delivered to the site.

One of the unique aspects of this system is the order of construction. The construction starts with the drilling of the shafts at the design locations. However, the drilled shafts are not filled with concrete immediately. Instead, plastic pier formers (see following section) are inserted into the top of the shafts. At this point, using a crane and two workers, the precast grade beams are set on blocks that have been laid out along the perimeter of the building. Thus using opposed wooden wedges, the grade beams are leveled and their positions determined. Steel reinforcement for the shafts typically consists of 1/2-inch-diameter, high-strength threaded rod. The threaded rod is inserted through a predrilled hole in the grade beam and lowered into the shaft, hanging from the grade beam. Concrete is then placed in the shafts and poured right up to the level of the grade beams, utilizing the plastic pier formers. The following day the blocks beneath the grade beams and the leveling wedges are removed. Raised wooden floors are constructed with the majority of the precast grade beam systems. Figures 11.1 and 11.2 show different stages of the precast grade beam construction process. Figure 11.1 shows the grade

Figure 11.1 Precast grade beam system. (Photograph courtesy of Bayshore Systems, Incorporated, Benicia, California.)

beams in place on blocks and leveled. A leveling wedge can be seen beneath the grade beam in the foreground of the photograph. The drilled shafts have already been drilled at this stage and plastic pier formers can be seen beneath the grade beams. Figure 11.2 shows the concrete being placed in one of the drilled shafts. The concrete truck can be seen in the background, pumping the concrete through sections of pipe extended through the air.

The main advantage of this system is the speed with which it can be constructed. It takes approximately eight man-hours to place the grade beams on the blocks, level them, and pour the drilled shafts for a typical residential structure. The beams are mass-produced, keeping their cost low. Additional time is saved in the field by eliminating the need to set and later remove forms for the grade beams. The time savings in the field renders the precast grade beam system competitive with the average post-tensioned slab system designed in the San Francisco Bay area—including the cost of the lumber and labor needed to construct the raised floor.

Using this system, over 30,000 homes have been constructed by two different companies in northern California. The majority of those

Figure 11.2 Concrete being placed in a drilled shaft beneath precast grade beams that are temporarily supported on wooden blocks. (Photograph courtesy of Bayshore Systems, Incorporated, Benicia, California.)

have been installed at sites with expansive soils or sloping ground. According to these companies, the only structures that have had any problems are those which have been situated on hillsides that have undergone deep rotational, global failures.

Plastic Pier Forms

These devices are used with drilled shaft systems during the placement of the concrete. They are plastic molded, tapered tubes or pliable plastic sheets bent around to rim the top of the shaft hole. Their purpose is to extend the drilled shaft up to the level of the grade beam. This is essential for the precast grade beam system described above. However, they also help reduce the potential for mushrooming due to overpouring the shaft. As discussed in Chapter 4, it is crucial to avoid mushrooming when constructing in areas with expansive soils.

The plastic forms also help to keep loose surface materials from falling into the hole. One disadvantage of this method is that although they are brightly colored as a warning, the forms do not cover the shafts before the concrete pour to keep people from stepping into them.

The idea of extending the drilled shaft above the ground surface is not a new one. Cardboard tubes have been used for this purpose for many years. However, the plastic pier form is more versatile than a cardboard tube. Because of its pliable nature, it can fit securely into any size shaft close to the design diameter.

A similar application of this type of material is to use it at a site with expansive soils to reduce the friction between the soil and the shaft in the zone of seasonal moisture variation. Reducing the friction will reduce the magnitude of the potential uplift force on the shaft due to expansion of the adjacent soil. To accomplish this, the plastic liner is inserted into the shaft past the suspected zone of seasonal moisture variation. The friction force is reduced because the coefficient of friction between the soil and the plastic liner is lower than that between the concrete and soil. Some feel that the liners become cumbersome when they are long enough to be inserted past the depth of moisture variation. The added labor and cost of the material itself may not offset the cost of extending the shaft several more feet into the soil to counter the potential uplift forces. Research is currently being conducted at the University of Washington utilizing different liner materials to determine the optimal material and to quantify any cost reductions and performance enhancements.

Speed Drill Rigs

Speed rigs consist of a truck-mounted drill rig that can be fully operated from the operator's seat, which is mounted in the back of the truck adjacent to the drilling boom. The gear shift, brakes, steering and acceleration of the truck can be operated from the rear of the vehicle using mechanical or hydraulic means. Although these rigs have been in operation since the 1960s, they have remained geographically isolated to the San Francisco Bay area and thus merit mention in this chapter. The speed rigs were originally designed as mechanical rigs mounted on 3/4- or 1-ton trucks with a drilling stroke of 6 feet. Presently, hydraulic speed rigs are mounted on 1- and 2-ton trucks with 14-to 16-feet drilling strokes. Figure 11.3 shows an example of a speed rig mounted on a 2-ton truck.

The advantage of speed rigs is that drilled shafts can be excavated very rapidly. A 15-foot shaft will typically take 1 to 2 minutes to excavate in stiff clay soils. The operator then drives the truck into position to drill

Figure 11.3 Typical speed drill rig. (Photograph courtesy of Bayshore Systems, Incorporated, Benicia, California.)

the next shaft without getting out of the seat. The drilling helper places a cover on the shaft to keep cuttings from falling into the shaft. A sweep can be attached to the auger flight to brush away the cuttings near the surface during the drilling. Figure 11.4 shows a completed shaft that was excavated using a sweep attached to the auger flight. The total time be-

Figure 11.4 Soil cuttings removed from the vicinity of a drilled shaft using a sweep. (Photograph courtesy of Bayshore Systems, Incorporated, Benicia, California.)

tween the start of two consecutive shafts can be as little as 2 to 3 minutes. Costs of drilling a drilled shaft foundation using these rigs on a typical residential tract home are estimated to be as low as $120.

The act of driving a truck from the rear and positioning it accurately over the marked spot for the drilled shaft sounds like a difficult task. However, with minimal practice, the operator can maneuver the rig as easily from the back as from inside the truck. In fact, because the operator is sitting at the back of the rig, the operator does not need someone behind the truck to help position the truck over the drilled shaft location.

Foundation Jacks

This innovation consists of installing jacks into the foundation system during the initial construction. Levels can be installed to monitor any differential movement of the foundation. If the differential movement reaches a specified level, the jacks are manually raised or lowered (depending on the direction of the movement and the play of the jacks) to compensate for this movement.

To the best of this author's knowledge this system is still in its infancy. It has not become commonly implemented owing to increased upfront costs. The system may also be subject to difficult maintenance in the event of a jammed jack.

This system also relies on the diligence of the owner periodically to monitor the sensors that detect the differential movement. The owner must also make adjustments to the jacks, which may not be practical for the average owner.

Moisture Detection and Monitoring Systems

This idea is similar to the foundation jacks described above in that the underlying principle is to monitor changes in the subgrade soils and alter them to stabilize the foundation. This system attempts to keep the subgrade moisture content relatively stable. This is important in minimizing volume changes of expansive soils. Moisture sensors are permanently inserted into the subgrade soils. Tubing connected to a water source is also installed beneath the foundation at strategic locations. If the sensors detect that the soil is beginning to dry out in specific areas, water is directed to that location. In this manner the subgrade moisture level is kept relatively high, keeping the clay in a more or less steady state.

This system is also in its infancy. Foreseeable disadvantages include high maintenance and equipment costs and clogging of the water input points. This system also relies on the owner to monitor the devices. Computerized monitoring and water input systems would eliminate the need for periodic monitoring but would also increase the cost.

Steel Helical Anchors

Steel helical anchors have been available for over 25 years. Their main use until recently has been to anchor utility poles and transmission towers. Recently they have become popular in repair applications for lightly loaded structures. Even more recently they have been utilized as the principal foundation system for new lightly loaded construction (A. B. Chance, 1988).

The steel helical anchors used for lightly loaded construction typically consist of a 5- to 10-foot length of 1-1/2- to 2-inch-square solid steel or 3- to 8-inch-diameter shaft with one or more 6- to 14-inch-diameter helices welded perpendicular to the length of steel. Shafts are manufactured with more than one helix to increase their capacity; however, a single helix is usually adequate for lightly loaded structures. Figure 11.5 illustrates a typical anchor and a lightly loaded application. A void box is shown in this example to illustrate its placement for a site with expansive soil.

The anchor is installed by attaching a rotary-drive motor to the top of the steel shaft and rotating the anchor into the ground. The motor can also be attached to limited access equipment, such as a bobcat or a portable drill rig, to facilitate the installation. The motor maintains a relatively slow speed of approximately 10 to 15 revolutions per minute to minimize the amount of disturbance to the soil while the anchor is being "screwed" into the ground. Once the length of shaft is completely rotated into the ground, another extension is fit over the top of the first section in the ground and is secured using bolts. The entire assembly is then advanced deeper into the soil using the rotary installer.

The purpose of the helix is to create a bearing surface (or uplift resistance in the case of expansive soil) to transfer the foundation load to the soil. The pitch of the helix is designed to minimize the amount of disturbance to the soil during its installation.

The anchor assembly is advanced into the soil until the bearing stratum is reached. The depth to the bearing stratum is estimated based on

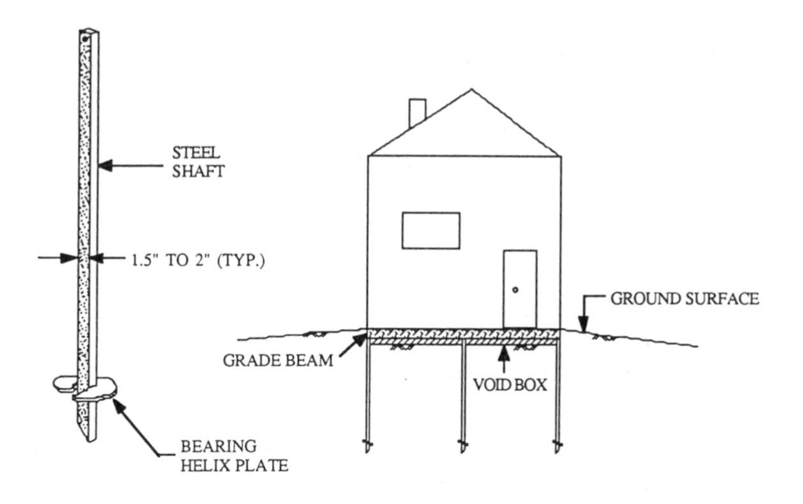

Figure 11.5 Typical helical steel anchor and application.

information presented in the soil investigation report. However, the drive assembly has a device attached to it that measures the torque necessary to advance the anchor at a constant rate into the soil. Load tests have been conducted to correlate the installation torque with soil bearing values (A. B. Chance, 1988). With this information, field personnel continue to advance the anchors until the design bearing value is achieved. It is necessary to be aware of what other parameters are affecting the depth of embedment. For example, it may appear that an anchor in a stiff clay has achieved adequate bearing near the surface, but if the clay is expansive it will be necessary to advance the anchor beneath the expected zone of moisture variation.

Product literature states that the anchors perform as well in tension as in compression (A. B. Chance, 1988). Therefore, they are suitable for both uplift and bearing applications. For example, they can be used to transfer a foundation load through a loose uncontrolled fill or through a soft compressible clay to a suitable bearing stratum. In this case the anchors would be in compression. Alternatively, the anchors could be used at a site with expansive soil to anchor the structure in a stratum below the zone of seasonal moisture variation.

The advantage of the anchors is their relatively narrow shafts. The limited surface area of the shafts reduces the amount of force that may be transferred from the soil to the anchor. For example, an anchor that has been installed in an expansive soil will be subject to a relatively small uplift load due to the limited surface area of the shaft with which the expansive soil is in contact. Similarly, an anchor installed through highly compressible soils will be subject to reduced downdrag forces.

Once the anchors are installed into the ground, a conventional grade beam system can be cast to transfer the wall loads into the anchors. Conventional reinforcing bars, or a fabricated termination, can be used at the top of the last extension to create a good bond between the grade beam and the anchor shaft.

Mini-Post Grouted Pile

Mini-piles consist of a small-diameter drilled hole that is pressure grouted and reinforced with a full-length, threaded reinforcing bar. The piles have the option to be post grouted as well. The piles gain their support from the soil via skin friction. The pressure grouting (and post-grouting) enhance the bond between the grout and the soil, thus increasing the skin friction and load carrying capacity.

The method of construction of the mini-pile is illustrated in Figure 11.6. The first step is to drill a small-diameter (typically 5 to 6 inches) borehole and then install a full-length casing into it. Next, the threaded bar (typically ranging from a No. 11 to No. 18) is installed in sections and attached with a coupler. The pressure grouting tube is also installed at this time. Spacers are periodically threaded onto the bar

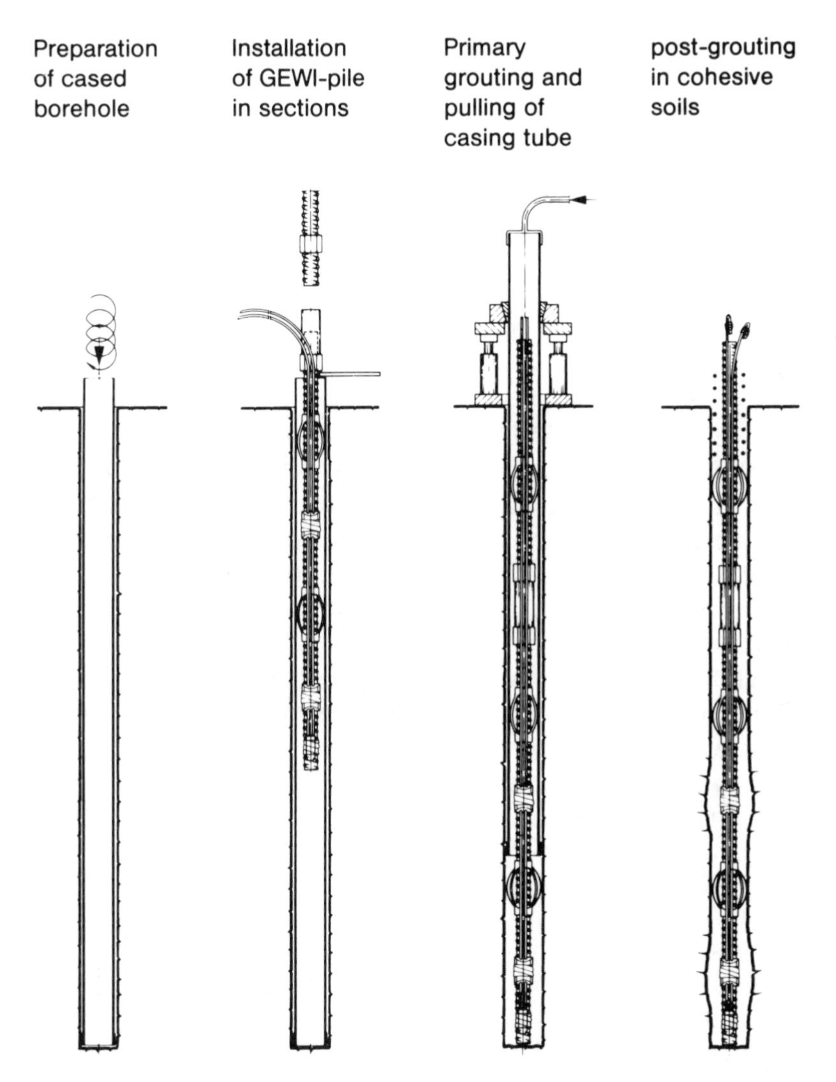

Figure 11.6 Mini-pile (GEWI-pile) installation. (Reprinted by permission of DYWIDAG Systems International.)

to ensure that the bar is centered in the borehole. Once this is done, the pressure grouting is conducted as the casing is withdrawn. If a higher skin friction is desired, the pile can then be post-grouted. This is especially important for clayey soils because the drilling operation can often "smooth" the sides of a borehole through clayey soil and significantly reduce skin-friction values. The post-grouting is typically conducted the day after the initial grouting and consists of a second application of grout under high pressure. The partially hardened, first application of grout is broken and forced deeper into the soil once the post-grouting begins, thus ensuring a better bond with the soil. However, if the soil is a relatively soft, clayey soil, post-grouting may not necessarily increase the capacity of the mini-pile because the potential for a shear failure to occur within the clay adjacent to the mini-pile still exists.

The slenderness of the mini-pile does create a potential for the piles to buckle. The stability of the piles should be checked in terms of buckling for each application. In most cases the surrounding soil will be sufficiently strong to keep the pile from buckling. Sites with very soft clays or peats may not be strong enough to support the pile from buckling, especially if the construction is located near a slope.

The small diameter of the drill hole enables high-performance rigs to drill more readily through boulders, rock, and hard substrata. The short sections of the threaded bar increase versatility especially in limited-access locations. Also, the threaded bar lends itself to easy connections for both load testing and permanent connection to the grade beam. Three methods of attaching the mini-pile to the structure are: (1) to bond it with the concrete of the foundation, (2) to use end anchorage by connecting an anchor-plate-nut assembly to the end of the threadbar, and (3) to use standard plate and nut connections (DYWIDAG, undated).

This system is gaining popularity, especially for upscale residential construction on hillsides. A contractor in Southern California who installs mini-piles views them as one of the up-and-coming foundation systems for the following reasons:

1. increasingly stringent building codes near hillsides that make deep foundations, in general, more economically feasible;
2. the relative ease at which the small-diameter holes can be bored through rocks and boulders;

3. the increased need for systems that can be used in sites with poor access such as hillsides or limited overhead clearance, and the increased use of sites with challenging soil conditions that may require a more innovative foundation solution; and
4. the ease with which mini-piles can be load-tested if installers or inspectors are uncomfortable with the piles' assumed capacity.

Fibrous Concrete Reinforcing

Fibrous concrete reinforcing is a product that is added to concrete to increase its ductility and its resistance to shrinkage cracks. Shrinkage cracks are an important issue for slab construction. Fibers made of high-tensile-strength polypropylene, approximately 3/4 inches to 1 inch long, are added to the concrete mix during the initial mixing. The fibers are typically added to the concrete at a minimum rate of 1.5 pounds of fibers per cubic yard of concrete. The fibers are then distributed throughout the concrete during the mixing process.

The potential for shrinkage cracks is reduced because the fibers are able to span any localized area that is beginning to open up (crack) at the microscopic level. Because the fibers have a high tensile strength and because they are distributed throughout the concrete, they are better able to hold the concrete together than wire mesh.

The addition of synthetic fibers has an advantage over the more commonly used welded wire mesh for two reasons. First, it is much less labor-intensive. The fibers are added at the concrete mixing plant; therefore, during the pour in the field, there is no additional work to be done related to the inclusion of the fibers. Second, welded wire mesh is rarely placed in the proper position within the slab to be effective. The fibers are located throughout the slab, eliminating this potential problem.

Strictly from a materials standpoint, the addition of fibers adds approximately 10% to the cost of the concrete. The reduction in labor costs and in the potential for cracking (and potential lawsuits) typically offsets this cost in the long run.

Adding fibers does have one aesthetic problem, commonly referred to as "whiskers." This occurs at the exposed surface of the slab where the fibers are able to protrude from the concrete, creating a whisker effect. The current trend in the industry to combat this problem is to minimize the length and cross section of the fibers, yet still keep them large enough to be effective. With time, these protruding fibers wear down to the point where they are much less noticeable.

It is important to note that the addition of fibers adds negligible flexural strength to the concrete slab. If relatively large amounts of subgrade movement are expected, fibers cannot replace the effectiveness of reinforcing steel in the slab.

Artificial Polystyrene Fills

Artificial fills typically consist of an expanded polystyrene material. The idea of using an artificial fill material has been around since the late 1960s. The advantage of using an artificial fill is that the weight of the fill material is much less than the weight of the compacted soil. The unit weight of expanded polystyrene fill may be as low as 3 pounds per cubic foot (pcf) (Monahan, 1986) compared to a unit weight of 100 to 130 pcf for compacted fill. This becomes very useful if the subgrade soils are highly compressible. The lighter-weight material adds less stress to underlying compressible soils, causing less settlement. The addition of expanded polystyrene fill at a site in order to make design grade will also reduce the negative skin friction on a driven pile or a drilled shaft foundation system.

Another use of artificial fills is as replacement material during a removal and replacement operation. This situation will actually create a weight reduction on the remaining soils. For example, a highly compressible clay may weigh approximately 100 pcf. Therefore, the depth of removal can be designed in such a manner that the stress applied to the remaining soil due to the construction of the building and the placement of the expanded polystyrene is equal to the stress that was released as a result of the removal of the compressible clay. Theoretically, this design would lead to negligible settlement.

As an example, the properties of a product used beneath the abutments at the Pickford Bridge crossing in Pickford, Michigan, in 1974 are listed below:

Compressive Strength & 5% deflection (ASTM D-1621-59T)	= 5000 psf
Water Absorption by Volume (ASTM C-272-53)	= 0.25%
Density	= 2.5 pcf

Another advantage of the expanded polystyrene fills is the versatility of the material. The expanded polystyrene can be delivered to the site as precast shapes or blocks of any size (Flaate, 1989) or the material can be pumped directly into the area requiring the fill. Additionally, the material can be fabricated with different compressive strengths and unit weights.

The material is not expected to decay with time because it is a very stable chemical compound (Flaate, 1989). According to Flaate, samples that have been tested from existing fills have actually shown increases in strength with time.

One disadvantage of the expanded polystyrene fills is that their light weight may require them to be held down when used below the water table, and their potential to absorb water may reduce the light-weight advantage. Additionally, exposure to artificial materials may cause the expanded polystyrene to dissolve. An example of this may be a spill of gasoline in the vicinity of a polystyrene fill. The cost of the polystyrene fills is approximately five times that of a compacted fill (Monahan, 1986); however, the time savings, material availability, and the reduction of potential settlement may make it possible to build at a site where it might otherwise have not been possible.

Because of the tremendous weight advantage of the polystyrene material, it is likely to become a common fill material in the future.

Laminated Polystyrene Forms

These forms for concrete are prefabricated out of polystyrene in the design shop. Once they are delivered to the site they are used to form up concrete walls. The main difference between these and conventional wood forms is that they remain in place as part of the structure.

Leaving the forms in place serves two purposes. The first is that the forms are waterproof and thus form a water-tight environment for the concrete to cure. This leads to higher-strength concrete. The second is that the forms provide additional insulation for the structure, which is helpful in cold weather regions.

SUMMARY

Presumably there are additional innovative products and systems being utilized within the lightly loaded foundation industry that have not been discussed in this text. The above innovations have been included as examples of some of the recent changes that the authors have encountered, and they are presented here to provide ideas for those involved in the lightly loaded foundation industry.

CLOSING COMMENTS

The purpose of this book is to expose the reader to a variety of different foundation solutions for lightly loaded structures founded upon challenging soil and site conditions. The key item to keep in mind is that there are many sites currently being developed that may not be suitable for the conventional lightly loaded foundation system of that region. Unfortunately, the nature of lightly loaded construction is such that the people involved in design and construction may become complacent. It is important to remember that simply because a building does not weigh as much as the 15-story office complex down the street, one should not assume that the potential for foundation movement is negligible.

> The least expensive foundation system is that which can be constructed with the lowest up-front costs, without sacrificing the long-term integrity of the structure.

Because of the dwindling number of "good" building sites and the increased liability involved with construction, it is strongly recommended that every building site be investigated by an experienced geotechnical engineer and the foundation and/or grading recommendations, resulting from the investigation, should be strictly adhered to.

An important concept to understand and remember when evaluating different foundation systems for challenging building sites is that in the long run, the most cost-effective system is not always the one that is least expensive to construct. The initial construction costs must be weighed against the likelihood that foundation remediation costs will be required in the future. Challenging building sites influence this comparison of up-front costs to remedial costs by increasing the likelihood of a foundation problem in the future. The least expensive foundation system is that which can be constructed with the lowest up-front costs, without sacrificing the long-term integrity of the structure.

RELATED REFERENCES

A. B. CHANCE COMPANY, *Products for the 80's: Foundations, Tiebacks, Pilings, Anchor Systems—Chance Foundation and Underpinning System,* 02150/CHA, BuyLine 1386, 1988.

COLEMAN, T. A., "Polystyrene Foam Is Competitive, Lightweight Fill," *Civil Engineering,* February 1974, pp. 68–69.

CONCO CEMENT COMPANY, *Conco Cement Company Brochure—Residential,* Concord, CA, undated.

DYWIDAG SYSTEMS INTERNATIONAL, *DYWIDAG GEWI-Pile Brochure,* Lemont, IL, undated.

FIBERMESH COMPANY, *Fibermesh®—Micro-Reinforcement System Brochure,* Chattanooga, TN, 1989.

FLAATE, K., "The (Geo)Technique of Superlight Materials," *The Art and Science of Geotechnical Engineering At the Dawn of the Twenty-First Century,* Prentice Hall, Englewood Cliffs, NJ, 1989.

FRYDENLUND, T. E., *Superlight Fill Materials,* Norwegian Road Research Laboratory, Publ. No. 60, pp. 11–14, 1986.

MONAHAN, E. J., *Construction Of and On Compacted Fills,* John Wiley & Sons, New York, 1986.

WISS, JANNEY, ELSTNER & ASSOCIATES, INC., *Static Load Test of Fibermesh® Versus Welded Wire Fabric,* F.E.D. Report No. 5, Fibermesh Company, Chattanooga, TN, 1986.

APPENDIX A

Example of a Foundation Inspection Checklist

Below is an example of a fairly rigorous checklist for the inspection of different foundation systems. This particular list has been adapted from the required inspection guidelines prepared by Home Buyers Warranty for sites with expansive soils. The Home Buyers Warranty is a company that provides insurance for builders against future building defects. One of the requirements of the insurance policy is to meet the following inspection requirements. The inspection is conducted by a representative of Home Buyers Warranty or by the on-site inspecting engineer.

The purpose of presenting this list is to provide an example of a checklist that may be applied to the reader's individual location. It is not intended to be a cure-all for all foundation problems. An inspection by a qualified, experienced inspector is still recommended for all sites. Items contained in Chapter 10 that are not listed here should also be considered during construction. This list is generally concerned with proper dimensions and construction and the quality of the materials. The condition of the subgrade and underlying soils should also be considered.

Inspection Item	Relevant Foundation Types	Inspection Specifics and Tolerances
Footing	Isolated Spread Footing Continuous Footing	Proper location—within 1/2″ from wall center and footing center unless engineer specifies other acceptable tolerance. Proper size—within 1″ in height (depth) and width. Proper key-in/wall tie. Proper steel grade and placement.
Foundation Drain (where required)	Isolated Spread Footing Continuous Footing Drilled Shafts Strutural Slab Post-Tensioned Slab	Drain material perforated; permits water to enter and discharge from system. Gravel/moisture barrier surrounds drain as per plans and specs. Proper drain placement.
Slab-on-Grade: Partition Void Framing (where required)	Isolated Spread Footing Continous Footing Drilled Shafts	Proper partition void; void installation as per plans and specs. Proper expansion joint around slab perimeter. Proper tooled/control joints within slab as per plans and specs. Stairways, landings, framing in general, constructed properly to allow slab to float as designed. Proper furnace duct sleeving. Column/pipe/conduit sleeving material and installation as per plans and specs.
Foundation Void (where required)	Isolated Spread Footing Drilled Shafts	Proper void size and material—correct height, width, and length. Void material in good condition—wet or collapsed void not permitted. Proper placement—against shafts, footings, and under all required walls. All concrete overpour to be cleaned away from void.
Grade Beam	Isolated Spread Footing Drilled Shafts	Proper wall dimensions—correct height, width, and length. Proper steel amounts and placement as per plans and specs.

Inspection Item	Relevant Foundation Types	Inspection Specifics and Tolerances
		Steel continuous. Steel size and grade correct. Proper placement on foundation supporting system: *Footing:* within 1/2″ from wall center and footing center. *Drilled shaft:* 50% off shaft center maximum; no two adjacent shafts to be out of alignment; 10% of shafts permitted to be at 50% off center tolerance.
Shaft Drilling	Drilled Shafts	Proper shaft size and depth as per plans and specs. Upper 3 feet of shaft bell—2″ maximum tolerance in excess of specified shaft diameter for drill wobble. Check groundwater conditions. Casing requirement—Greater than 6″ of water in hole or cave-in of shaft sides requires engineer recommendations on casing. Mushroomed shafts to be cleaned to specified shaft diameter. Within 1″ across shaft diameter maximum permitted. Proper shaft tie into wall. Steel projection as per plans and specs. Shaft drilled into [formation] or refusal. Engineer recommendations required on shafts drilled 20 feet or greater not encountering [formation] or refusal. Check for clean hole—2″ of fill maximum permitted [end bearing shafts only]. Proper shaft alignment—2% out of plumb maximum over entire length of shaft. Proper shaft location. Shafts to be located properly around foundation perimeter and under all proposed columns and supporting members as per plans and specs.

Inspection Item	Relevant Foundation Types	Inspection Specifics and Tolerances
Slab Steel/ Tendon Placement	Structural Slab Post-Tensioned Slab	Steel tied and lapped properly. Steel size and grade correct. Slab thickness correct. Beam widths, depths, spacing correct. Tendons placed properly and sheathed.
Tensioning	Post-Tensioned Slab	Adequate post-tensioning anchorage. All tendons are stressed in accordance with the design requirements.

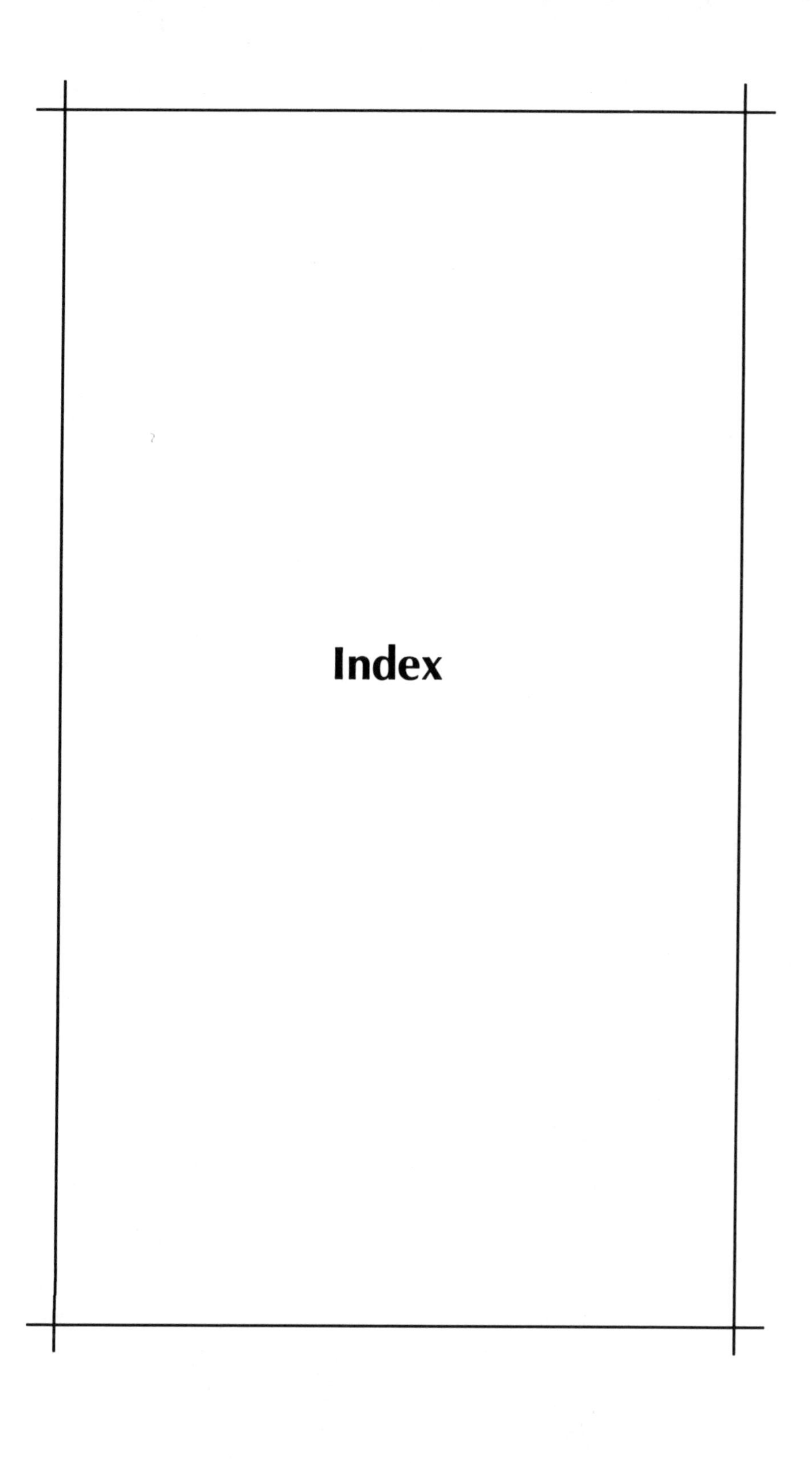

Index

A

B

C

D

E

F

G

H

I

P

Q

R

S

T

U

V

W